T0214263

The main theme of this book is the idea that quantum mechanics is valid not only for microscopic objects but also for the macroscopic apparatus used for quantum mechanical measurements. The author demonstrates the intimate relations between quantum mechanics and its interpretation that are induced by the quantum mechanical measurement process. Consequently, the book is concerned both with the philosophical, metatheoretical problems of interpretation and with the more formal problems of quantum object theory.

The consequences of this approach turn out to be partly very promising and partly rather disappointing. On the one hand, it is possible to give a rigorous justification of some important aspects of interpretation, such as probability, by means of object theory. On the other hand, the problem of the objectification of measurement results leads to inconsistencies that cannot be resolved in an obvious way. This open problem has far-reaching consequences for the possibility of recognising an objective reality in physics.

The book will be of interest to graduate students and researchers in physics, the philosophy of science, and philosophy.

THE INTERPRETATION OF QUANTUM MECHANICS
AND THE MEASUREMENT PROCESS

THE INTERPRETATION OF QUANTUM MECHANICS AND THE MEASUREMENT PROCESS

PETER MITTELSTAEDT

Institute for Theoretical Physics, University of Cologne, Germany

CAMBRIDGE
UNIVERSITY PRESS

PUBLISHED BY THE PRESS SYNDICATE OF THE UNIVERSITY OF CAMBRIDGE
The Pitt Building, Trumpington Street, Cambridge, United Kingdom

CAMBRIDGE UNIVERSITY PRESS
The Edinburgh Building, Cambridge CB2 2RU, UK
40 West 20th Street, New York NY 10011–4211, USA
477 Williamstown Road, Port Melbourne, VIC 3207, Australia
Ruiz de Alarcón 13, 28014 Madrid, Spain
Dock House, The Waterfront, Cape Town 8001, South Africa

http://www.cambridge.org

© Cambridge University Press 1998

This book is in copyright. Subject to statutory exception
and to the provisions of relevant collective licensing agreements,
no reproduction of any part may take place without
the written permission of Cambridge University Press.

First published 1998
First paperback edition 2004

A catalogue record for this book is available from the British Library

ISBN 0 521 55445 4 hardback
ISBN 0 521 60281 5 paperback

Contents

Preface

This book, on the interpretation of quantum mechanics and the measurement process, has evolved from lectures which I gave at the University of Turku (Finland) in 1991 and later in several improved and extended versions at the University of Cologne. In these lectures as well as in the present book I have aimed to show the intimate relations between quantum mechanics and its interpretation that are induced by the quantum mechanical measurement process. Consequently, the book is concerned both with the philosophical, metatheoretical problems of interpretation and with the more formal problems of quantum object theory.

The book is based on the idea that quantum mechanics is valid not only for microscopic objects but also for the macroscopic apparatus used for quantum mechanical measurements. We illustrate the consequences of this assumption, which turn out to be partly very promising and partly rather disappointing. On the one hand we can give a rigorous justification of some important parts of the interpretation, such as the probability interpretation, by means of object theory (chapter 3). On the other hand, the problem of the objectification of measurement results leads to inconsistencies that cannot be resolved in an obvious way (chapter 4). This open problem has far-reaching consequences for the possibility of recognising an objective reality in physics.

The manuscript of this book was carefully written in $\TeX$ by Dipl. Phys. Falko Spiller. In addition, he proposed numerous small corrections and improvements of the first version of the text. His helpful cooperation and his continued interest in the progress of this book are gratefully acknowledged.

Furthermore, I wish to express my gratitude to Dr. Julian Barbour for reading carefully the whole manuscript as a native English speaker and physicist. He proposed many changes and improvements of the language. Moreover, he made several interesting physical suggestions which are partly realized in the final version of the book.

Finally, I want to thank Dr Simon Capelin of Cambridge University Press for his encouragement to write this book and for his kind cooperation during the last two years.

Peter Mittelstaedt
Cologne

1

Introduction

1.1 Measurement-induced interrelations between quantum mechanics and its interpretation

1.1(a) The development of quantum mechanics

The formalism of quantum mechanics was developed within the very short period of a few months in 1925 and 1926 by Heisenberg [Heis 25] and by Schrödinger [Schrö 26], respectively. Together with the contributions of Born and Jordan [BoJo 26], [BHJ 26], Dirac [Dir 26] and others, the formalism of this theory was already brought in 1926 into its final form, which is still used in present-day text books. (For all details of the historical development, we refer to the monograph by M. Jammer [Jam 74].) It is a very remarkable fact that a theory which was formulated 70 years ago has never been corrected or improved and is still considered to be valid. Numerous experiments performed during this long period to test the theory have confirmed it to a very high degree of accuracy without any exception. Hence there are good reasons to believe that quantum mechanics is universally valid and can be applied to all domains of reality, i.e., to atoms, molecules, macroscopic bodies, and to the whole universe.

However, the interpretation of the new theory was at the time of its mathematical formulation still an almost open problem. Any interpretation of quantum theory should provide interrelations between the theoretical expressions of the theory and possible experimental outcomes. In particular, an interpretation of quantum mechanics has to clarify which are the theoretical terms that correspond to measurable quantities and whether there are limitations of the measurability, e.g. whether there is a limit to the simultaneous measurability of two observables. Another essential problem is the question of what kind of experimental results could correspond to the Schödinger wave function, which turns out to be a very important theoretical entity.

1

The first consistent and complete interpretation of quantum mechanics was formulated by Niels Bohr in 1927 in his Como lecture [Bohr 28] and was later called the *Copenhagen interpretation*. In this interpretation, Bohr made use of a methodological requirement that was first formulated by Einstein in his investigations of special and general relativity: measuring instruments that are used for the interpretation of theoretical expressions must be truly existing physical objects. For example, time intervals are measured by clocks whose mechanisms are subject to the laws of physics, and distances in space are measured by measuring rods that are not assumed to be ideally rigid bodies, which do not exist in nature. In this sense, Bohr always assumed that the apparatus for measuring observables like position, momentum, energy, etc. can actually be constructed in a laboratory.

By means of this methodological premise, Bohr could explain one of the most surprising features of the theory, which he called *complementarity*. In quantum mechanics, two observables A and B that are canonically conjugate in the sense of classical mechanics cannot be measured simultaneously. The most prominent example of this non-classical behaviour is the complementarity of position q and momentum p. Bohr explained the complementarity of the observables p and q in the following way: the measuring apparatuses[†] $M(p)$ and $M(q)$ that could be used for measuring p and q, respectively, are mutually exclusive. In other words, there is no real instrument $M(p, q)$ that could be used for a joint measurement of p and q. The second methodological premise that is used in the Copenhagen interpretation is the hypothesis of the classicality of measuring instruments. This means that the apparatuses that are used for testing quantum mechanics must not only truly exist in the sense of physics, but these apparatuses must also be macroscopic instruments that are subject to the laws of classical physics. Consequently, the experimental outcomes of measurements are events in the sense of classical physics and can be treated by means of classical theories like mechanics, electrodynamics, etc. In this way, the strange and paradoxical features of quantum mechanics disappear completely in the measurement results, which can thus be described by means of classical physics and ordinary language. The interrelations between quantum mechanics, its interpretation, and the measuring process within the framework of the Copenhagen interpretation are schematically shown in Fig. 1.1.

The fact that the observer of a quantum system is always 'on the safe side' and not affected by quantum paradoxes is expressed in the following statement, which was made by Bohr many years later in order to reject

---

† Here we use the form 'apparatuses' to distinguish unambiguously the plural from the singular.

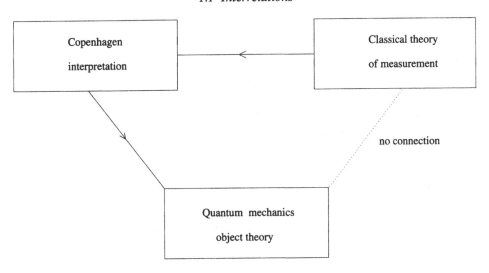

Fig. 1.1 Interrelations between quantum object theory, its interpretation, and the measuring process in the Copenhagen interpretation.

the attempt to modify even logic by the introduction of quantum logic: 'Incidentally, it would seem that the recourse to three-valued logic sometimes proposed as a means for dealing with the paradoxical features of quantum theory is not suited to give a clearer account of the situation, since all well-defined experimental evidence, even if it cannot be analysed in terms of classical physics, must be expressed in ordinary language making use of common logic' [Bohr 48]. (We add that all experimental evidence must also be expressed in terms of classical physics.) Although quantum logic is not the topic of the present book, we mention this statement here since it illustrates the Copenhagen interpretation in a very clear and convincing way.

Once the explanation of the Copenhagen interpretation for the impossibility of joint measurements of complementary observables had been asserted, one could try to question this explanation by constructing gedanken experiments that do allow the simultaneous measurement of position and momentum, say, of a quantum system. This was, indeed, the strategy by means of which Einstein tried to show that the claimed restrictions of joint measurements do not really exist. However, in spite of a large number of ingenious gedanken experiments proposed by Einstein, in every single case Bohr could show that the complementarity principle could not be circumvented by such new and sophisticated experiments. The whole debate between Bohr and Einstein is reported in an article by Niels Bohr written for the volume *Albert Einstein, Philosopher–Scientist* [Bohr 49].

1.1(b) Quantum theory of measurement

Einstein's requirement that measuring instruments should be real physical objects can be completely fulfilled in special and general relativity. Moreover, in these theories it is even possible to use instruments that are not only real in the sense of physics (in general) but also subject to the laws of relativity. This means that the theory can be verified or falsified, and interpreted, by means of measuring processes that are governed by the same physical laws that they should test. It is obvious that a theory of this kind must be rich enough that the processes needed for testing and interpreting the theory are contained within the domain of phenomena that are described by the theory. This metatheoretical property, which will be called *semantical completeness*, is actually given in special and general relativity. The formulation of a theory that contains the means of its own justification was not fully realized by Einstein himself. It was completed later by the construction of certain measuring devices using light rays and particle trajectories [KuHo 62], [MaWh 64] and by the formulation of an axiomatic system based on these instruments [EPS 72]. The requirement that the measuring processes which are used for the justification of a given theory are determined by the laws of the same theory was first applied to quantum mechanics by J. von Neumann [Neu 32]. In contrast to the Copenhagen interpretation, von Neumann treated the measuring process in quantum mechanics as a quantum mechanical process and the measuring apparatus as a proper quantum system. Consequently, in this theory of measurement, the object system S and the measuring apparatus M are both considered as quantum systems, the interaction of which is described by a quantum mechanical Hamilton operator $H(S + M)$, which acts on the compound system $S + M$. Roughly speaking this means that the measuring process is treated like a scattering process between the quantum systems S and M.

If this concept of measurement is accepted, the following problem arises. On the one hand, the measuring process serves as a means to justify or to falsify quantum mechanics (QM) and to provide an interpretation that relates the theoretical terms of the theory to experimental data. Hence the measuring process is part of a metatheory M(QM) that contains the semantics of the object theory in question, i.e., quantum mechanics. On the other hand, the measuring process is also a real physical process, and as a physical process it is subject to the laws of quantum mechanics. This means that the measuring process plays a twofold rôle: it serves as a means to interpret the quantum object theory, and it is also a real physical process that belongs to the domain of phenomena described by the object theory.

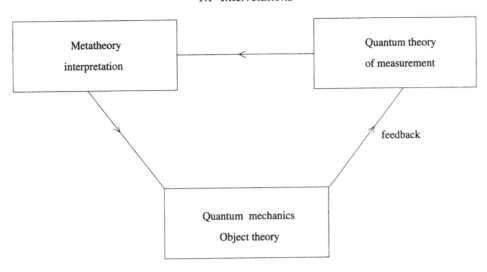

Fig. 1.2 Interrelations between quantum theory and its interpretation within the framework of a quantum theory of measurement.

The interrelations between a semantically complete quantum mechanics, its interpretation, and the quantum theory of measurement in the sense of von Neumann are schematically shown in Fig. 1.2. In contrast to the scheme of the Copenhagen interpretation (Fig. 1.1), the present scheme contains a 'feedback' from the object theory to the theory of measurement.

1.1(c) The twofold rôle of the measuring process

Whenever the measuring process may be considered as a quantum mechanical process, the quantum mechanical object theory and its metatheory consisting of the semantics and the interpretation of the object theory are connected by the measuring process in a twofold way. This is the content of the scheme shown in Fig. 1.2. From a methodological point of view, the measuring apparatuses do not belong to the domain of reality of the considered object theory but rather serve as means for establishing a semantics and an interpretation, which provides a relation between object-theoretical terms and experimental results. For this reason, the measuring apparatuses belong to the metatheory. On the other hand, if quantum theory is assumed to be semantically complete, then the measuring apparatuses, considered as physical objects, belong to the domain of reality of the quantum object theory and are subject to the laws of this theory.

The measurement-induced interrelations between quantum object theory and its interpretation, i.e., its metatheory, express a certain self-referentiality

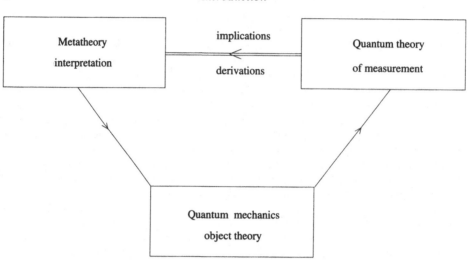

Fig. 1.3 Interrelations between quantum theory and its interpretation. The case of self-referential consistency.

of quantum mechanics. The interpretation of the theory is influenced by the properties of the measuring instruments, which are, considered as physical objects, subject to the laws of quantum object theory. Clearly, this way of reasoning presupposes that the physical processes which are used for the preparation of the object system and for measurements of the observable quantities are contained within the domain of phenomena that are described by the theory. If this requirement of 'semantic completeness' is fulfilled – and this is the case if quantum theory is considered as 'universally valid' – then two different situations could arise. First, it could happen that some parts of the interpretation are not independent requirements but derivable from quantum theory. Then parts of the interpretation and the semantics of truth for quantum mechanical propositions should also be fulfilled as a consequence of the theory itself. This situation is called here *self-referential consistency* and is shown schematically in Fig. 1.3. In addition to the general interrelations between object theory and metatheory which appear whenever a quantum theory of measurement is used and which are shown in Fig. 1.2, we have here implications from the theory of measurement for the interpretation such that parts of the interpretation can be derived. An interesting example of self-referential consistency will be discussed within the framework of the statistical interpretation of quantum mechanics in chapter 3.

However, it could also happen that the quantum object theory contradicts some parts of the interpretation and the corresponding underlying preconditions. Such *self-referential inconsistency* indicates a strong semantic

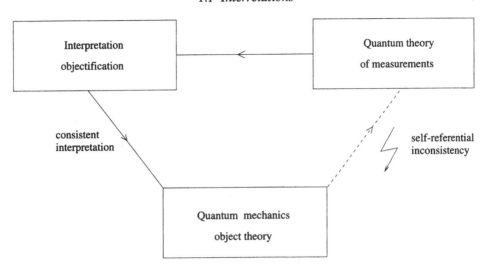

Fig. 1.4 Interrelations between quantum mechanics and its interpretation. The case of self-referential inconsistency.

inconsistency of the theory. A self-referential inconsistency of this kind, concerning the problem of objectification, is investigated in chapter 4.

Even if the object theory is semantically complete in the sense of Fig. 1.2 and self-consistent, it could still happen that the observer plus apparatus M is contained in the object system S as a subsystem: $M \subset S$. At first glance, this situation looks rather artificial. However, within the framework of quantum cosmology it is obvious that the apparatus and the observer are parts of the object system, which, in this case, is identical with the entire universe (Fig. 1.5). It turns out that in this extreme situation the possibilities of measurement, which means measurement from inside, are strongly restricted compared to the usual situation of measurements from outside. These restrictions, which will be discussed in chapter 5, have interesting consequences for the metatheory and the various interpretations of quantum mechanics.

Even if quantum object theory were semantically complete and without self-referential inconsistencies, the self-referential character of the theory could lead to serious methodological problems. Self-referentiality induces a logical situation similar to that discussed in the famous investigations of Gödel and Tarski. These problems have been mentioned by several authors (e.g. [DaCh 77]; [PeZu 82]) and deserve to be taken seriously. We shall discuss some questions and consequences of the self-referentiality of quantum mechanics in chapter 5. There are indeed some interesting results. However, these metalogical problems still require further elaboration into a

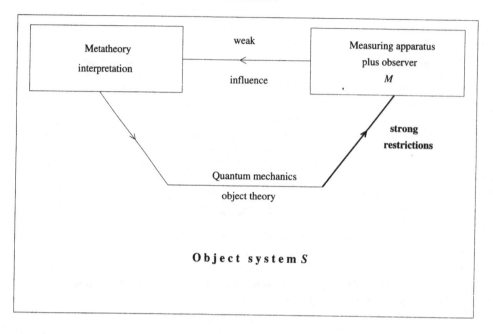

Fig. 1.5 Interrelations between quantum mechanics and its interpretation. Measurements from inside when $M \subset S$.

rigorous formalization before their implications can properly be discussed. The problems that arise from the self-referential character of the theory should appear within the framework of a formal language of quantum physics, provided the measuring process has been incorporated into the language. However, this incorporation has not yet been achieved.

1.2 Interpretations of quantum mechanics

The methodological aspects that we have just discussed will be applied to several interpretations and further investigated in the following chapters. As a preparation for these considerations, we shall describe here briefly three interpretations of quantum mechanics that are probably the most important ones. It is not claimed here that this short review is exhaustive, but we think that the most interesting aspects and problems have been taken into account. In particular we will not discuss here the various versions of the so-called modal interpretation, for the following reasons. First, although there is a lively discussion of this interpretation in the present literature, up to now there has been no general agreement about the value, the usefulness, and the philosophical implications of these approaches [Bub 92,94], [Die 89,93], [Koch 85], [Fraa 91]. Second, for an exhaustive comparison and evaluation of the different versions of the modal interpretation, one needs advanced mathematical tools that are beyond the scope of the present book. For more details, we refer to the investigations of Cassinelli and Lahti [CaLa 93,95]. Third, it is the common aim of the various modal interpretations to restore objectivity in quantum mechanics as far as possible. However, on account of the nonobjectification theorems that will be discussed in chapter 4, this goal can be achieved only contextually, i.e., with a dependence on the observables and the state.

1.2(a) The minimal interpretation

The minimal interpretation is the weakest interpretation of the quantum mechanical formalism if one goes one step beyond the Copenhagen interpretation but preserves the requirement that any measurement leads to a well-defined result. In contrast to the Copenhagen interpretation, the minimal interpretation does not assume that measuring instruments are macroscopic bodies subject to the laws of classical physics. Instead, measuring apparatuses are considered as proper quantum systems and treated by means of quantum mechanics. This means that with respect to measuring instruments the minimal interpretation replaces Bohr's position by von Neumann's approach.

On the other hand, the minimal interpretation preserves the empiristic and positivistic attitude, in the sense of David Hume [Hume 1739] and of Ernst Mach [Mach 26] respectively, of the Copenhagen interpretation. Indeed, it avoids statements about object systems and their properties and instead refers to observed data only. Since 'observations' in quantum mechanics are always the last step in a measuring process, the 'observed data' are merely the values of a 'pointer' of a measuring apparatus. For this reason, 'pointer values' play an important rôle in the minimal interpretation.

As already mentioned, the minimal interpretation also adopts another requirement of the Copenhagen interpretation which is self-evident in the latter interpretation and not explicitly mentioned: after the measuring process, the pointer of the apparatus should have a well-defined value that represents the measurement result. For a classical apparatus, this assumption of 'pointer objectification' is obviously fulfilled, and also for a macroscopic quantum apparatus it was considered for a long time not to be associated with any serious problems. We mention this requirement here explicitly, since during the last decade it has become clear that the evolution of objective pointer values in a quantum mechanical measuring process is not yet properly understood. These questions will be discussed in full detail in chapter 4.

There is one situation in which the positivistic attitude of the minimal interpretation is suspended. Before a measuring apparatus can be applied, one has to make sure that it is really convenient for measuring an observable A, say. This means that one has first to calibrate the measuring apparatus in such a way that a pointer value Z_i corresponds to a well-defined value A_i of the measured observable. If, for example, one wants to calibrate a weighing-machine, one must put a body of weight w_i, say, on the apparatus and define the scale Z of the pointer such that the measurement result w_i is indicated by a pointer value Z_i. If this method is extended to other values of Z and w, one finally arrives at some *pointer function* $w_i = f(Z_i)$.

Precisely this procedure is also applied in the minimal interpretation. Assume that there is an object system with the value A_i of the observable A. (This is the assumption that suspends the empiricistic position.) A measuring process that is suitable for the observable A should then lead to a pointer value Z_i such that the measurement result $A_i = f(Z_i)$ is indicated by the value Z_i and a pointer function f that depends on the construction of the apparatus. We will not discuss here the question of the way in which a quantum system can be prepared such that it possesses a definite value A_i of the observable A. Instead, we simply assume that a quantum object system with this property is given. This assumption is not a vague hypothesis, since in many interesting cases there are known preparation methods that lead to systems with well-defined properties.

The calibration postulate is concerned with individual systems that are prepared in such a way that they possess a definite property A before the measurement. The postulate demands that this property can be verified by measurement with certainty, such that the result A of a measurement is indicated by a well-defined pointer value Z_A. However, at this stage of the discussion nothing is known about further properties B, C, ... of systems that are not compatible with the preparation property A. We know from the Copenhagen interpretation that in spite of the incommensurability of B and A, say, the property B can be tested by measurement but that the result of this experimental test is unpredictable.

It was first observed by Max Born [Born 26] that the preparation state φ^A of the system also provides some information about a property B that is incommensurable with the preparation property A. Indeed, the preparation state φ^A and the measured observable B provide a probability measure $p(\varphi^A, B)$, which fulfils the well-known Kolmogorov axioms of probability theory. The interpretation of this formal probability was first given by Born: the probability distribution $p(\varphi^A, B)$ is reproduced in the statistics of the measurement outcomes of B-tests that are performed on a large number of equally prepared systems.

For the sake of historical correctness it should be mentioned that this 'reproduction of the probability distribution in the statistics of measurement results', which is often called the Born interpretation, cannot be found in Born's early writings. Born assumed that the formal expression $p(\varphi^A, B)$ can be interpreted as the probability (in the sense of subjective ignorance) for the property B to pertain or not to pertain to the system. (For details, cf. [Jam 74] pp. 38ff.) This original Born interpretation, which was formulated for scattering processes, turned out, however, not to be tenable in the general case for two reasons. First, if the same way of reasoning is applied to double-slit experiments with interference structure, then some theoretical predictions are not in accordance with the experimental outcomes. (Cf. the discussion of this problem in subsection 4.2(b).) Second, in the general case it will also not be possible to relate the probabilities $p(\varphi^A, B)$ to properties (B or $\neg B$) of the object system, since this system will be disturbed by the measuring process or even destroyed. Hence in general the probabilities can be related only to the outcomes of the measuring process, i.e., to the pointer values after the measurement.

In order to conclude this subsection, we summarize its content by three postulates that characterize the *minimal interpretation* I_M. (In the spirit of this introductory chapter we formulate these postulates without mathematical details. A more technical presentation of the postulates will be given in the following chapters.)

1. *The calibration postulate (C_M).* If a quantum system is prepared in a state φ^A such that it possesses the property A, then a measurement of A must lead with certainty to a pointer value Z_A that indicates the result $A = f(Z_A)$ of the measuring process, where f is a convenient pointer function.

2. *The pointer objectification postulate (PO).* If a quantum system is prepared in an arbitrary state φ which does not allow prediction of the result of an A-measurement, then a measurement of A must lead to a well-defined (objective) pointer value, Z_A or $Z_{\neg A}$, which indicates that the property $A = f(Z_A)$ does or does not pertain to the object system. Here, however, the objectivity is only postulated for the pointer values and not necessarily for the corresponding system properties.

3. *The probability reproducibility condition (PR).* The probability distribution $p(\varphi, A_i)$ that is induced by the preparation state φ of the object system and the measured observable A with values A_i must be reproduced in the statistics of the pointer values $Z_i = f^{-1}(A_i)$ after measurements of A on a large number of equally prepared systems.

1.2(b) The realistic interpretation

The realistic interpretation differs from the Copenhagen interpretation in two respects. As in the minimal interpretation, the measuring instruments are not assumed to be classical instruments but proper quantum systems, and the interpretation is concerned not only with the measurement outcomes but also with the properties of an individual system. Hence, the realistic interpretation is of higher explanatory power than the minimal interpretation. It relates the theoretical expressions of quantum mechanics not only to the pointer values but also to the values of observables of the object system. The attribute 'realistic' is used here in order to indicate that the interpretation is concerned with the reality of object systems and their properties.

It is obvious that this stronger interpretation is not applicable in the general case. Indeed, the realistic interpretation presupposes that the measuring process avoids any unnecessary disturbance of the object system. In particular, if an object system is prepared in such a way that some value A_k of an observable A pertains to the system, then a subsequent measurement of A should preserve this value A_k. In this special case, any disturbance of the object system is unnecessary, since the system possesses already a value of the measured observable A. Within the systematics of premeasurement, discussed in chapter 2, measurements of this kind are called *repeatable*. If such a measurement is repeated several times, the result of these measurements remains unchanged.

On the basis of these explanations one can now formulate the main requirements of the realistic interpretation. In order to demonstrate clearly the similarities of the realistic and the minimal interpretation, and also the differences between them, we use the same terminology for characterizing the two interpretations. The first step is in both cases the calibration of the measuring apparatus. The calibration requirement of the realistic interpretation shows in particular that this interpretation is restricted to repeatable measurements. In the premise of the calibration postulate, one assumes – as in the minimal interpretation – that the individual object system is prepared in such a way that a value A_k of an observable A pertains to the system. In distinction to the minimal interpretation, in which this premise does not agree with the empiricistic attitude of the interpretation, it is here in complete accordance with all other assumptions, since the realistic interpretation is generally concerned with object systems and their properties.

The calibration postulate of the realistic interpretation, which is concerned with pointer values and with object values, can now be formulated in the following way. If an object system is prepared in a state φ^{A_i} such that the observable A possesses the value A_i, then a repeatable measurement of A must lead with certainty to a pointer value Z_i that indicates the result $A_i = f(Z_i)$, where f is a convenient pointer function. Furthermore, this measuring process must also lead to the value A_i of the object observable A. The Z-value Z_i of the pointer and the A-value A_i of the object system are connected by a pointer function f such that $A_i = f(Z_i)$.

The second requirement of the realistic interpretation is – as in the minimal interpretation – concerned with the general situation of an arbitrary preparation state φ. Even if in this case the result of an A-measurement cannot be predicted, the minimal interpretation assumes that the pointer of the measuring apparatus possesses some objective value Z_i indicating the measurement result $A_i = f(Z_i)$. The realistic interpretation postulates in addition that the object system also assumes an objective value A_i of the object observable A. This requirement is an extension of the realistic calibration postulate to the case of an arbitrary preparation. It will be denoted here as *system objectification*.

The third and last requirement of the realistic interpretation consists of another extension of the realistic calibration postulate. If a system is prepared in a state φ^{A_i} such that the system possesses the value A_i of A, then a subsequent measurement of A will lead with certainty to the value A_i of the system. In case of an arbitrary preparation, the system objectification postulate requires that the observable A will have some objective value, but no prediction can be made for the A-value of the system after the measurement. However, one knows from the minimal interpretation that the probability measure $p(\varphi, A_i)$ which is induced by the preparation φ and the measured observable A is reproduced in the statistics of the pointer values Z_i that indicate the measurement results A_i. In addition, the realistic interpretation postulates that the probability measure $p(\varphi, A_i)$ is also reproduced in the statistics of the post-measurement system values $A_i = f(Z_i)$. It is obvious that this requirement follows from the probability reproducibility condition (PR) of the minimal interpretation if in addition system objectification is presupposed. The three postulates of the realistic interpretation will, however, be treated here as independent requirements.

In conclusion, we summarize the content of this subsection by formulating three postulates that characterize the *realistic interpretation* I_R.

1. The calibration postulate (C_R). If a quantum system is prepared in a state φ^A such that it possesses the property A, then a measurement of A leads with certainty to a system state with the property A and to a pointer value Z_A that indicates the result A.

2. The system objectification postulate (SO). If a quantum system is prepared in an arbitrary state φ, then a measurement of A leads to an objective pointer value Z_i indicating the measurement result A_i. Moreover, the value A_i of A pertains actually to the object system after the measurement.

3. The probability reproducibility condition (SR). The probability distribution $p(\varphi, A_i)$ that is induced by the preparation φ and the measured observable A with values A_i is not only reproduced in the statistics of the pointer values Z_i that indicate the measurement results A_i but also in the statistics of the post-measurement system values $A_i = f(Z_i)$.

1.2(c) The many-worlds interpretation

The many-worlds interpretation was formulated by Everett [Eve 57] and Wheeler [Whe 57] in 1957 and was initially called the 'relative-state interpretation'. More than a decade later, this interpretation was elaborated and extended by many authors, in particular by DeWitt [DeW 71] and by Graham [Gra 70], and called the 'many-worlds interpretation'. The most important contributions to this interpretation can be found in a collection of papers edited by DeWitt and Graham [DeWG 73].

Like the minimal interpretation and the realistic interpretation, the many-worlds interpretation considers the formulation of quantum mechanics as sufficient for the description of the object system, the apparatus, and the measuring process. This means that here too it is unnecessary to refer to classical physics as a methodological background of quantum mechanics. In contrast to the Copenhagen interpretation, quantum physics is considered to be universal, i.e., applicable to microscopic systems, to macroscopic measuring instruments, to the human observer, and to the entire universe. With respect to these assumptions, the many-worlds interpretation agrees with the two interpretations discussed in subsections 1.2(a) and 1.2(b) above.

The many-worlds interpretation was not conceived as a new interpretation making new hypothetical assertions about the meaning of quantum mechanical terms. Instead this interpretation avoids any additional assumption that goes beyond the pure formalism, even the very few and weak assumptions that are made in the minimal interpretation. Hence, the many-worlds interpretation should be considered as the very interpretation of quantum mechanics – as something that can be read off from the formalism itself. At first glance, one may wonder whether under these restrictive conditions an interpretation can be formulated at all. The interesting and surprising result of the investigations mentioned is that a consistent interpretation of quantum mechanics can actually be found in this way.

The first step towards the many-worlds interpretation is not different from that for the two other interpretations mentioned. One has to define what kind of apparatus can be used for measuring a given observable A. As in the preceding sections this can be achieved by an implicit definition which is expressed by a calibration postulate. The weakest formulation is given again by the calibration postulate of the minimal interpretation, which will be denoted here as $(C_M)_{MW}$. It states that whenever an object system is prepared in a state φ_i^A such that the observable A possesses the value A_i then a repeatable measurement of A must lead with certainty to a pointer value Z_i that indicates the result $A_i = f(Z_i)$. In the present case, this result means that the object system actually possesses the A-value A_i after the measuring process. The Z-value Z_i and the A-value A_i after the measurement are connected by the pointer function f. In order to avoid unnecessary complications that have nothing to do with the present problem, we will confine our considerations here to the realistic version of the many-worlds interpretation, the first requirement of which is the realistic calibration postulate $(C_R)_{MW}$.

If the object system is prepared in an arbitrary state φ, the result of an A-measurement can no longer be predicted with certainty. In contrast to the realistic and minimal interpretations it is *not* assumed here that the pointer of the measuring apparatus possesses an objective value Z_i indicating that the object system possesses the objective value $A_i = f(A_i)$ of the observable A. These objectification requirements are additional assumptions that are beyond the limits of quantum mechanics. It will be shown in chapter 4, that pointer objectification as well as system objectification are assumptions that are *not* compatible with the formal results of quantum mechanics. For this reason, the many-worlds interpretation dispenses with these hypothetical assumptions. Instead, it follows strictly the quantum theory of measurement and tries to read off from the theory how the apparatus and the object system look after the measurement.

According to the quantum theory of measurement, which will be discussed in chapter 2, after the measuring process the total system $S + M$ is in a highly entangled pure state $\Psi(S + M)$. This means that the object system S and the apparatus M are in correlated mixed states W_S and W_M, respectively. In the simplest situation, a repeatable measurement of a discrete nondegenerate observable A, the mixed state of the object system consists of pure states φ^{A_i} corresponding to values A_i, with weights $p(\varphi, A_i)$ that are given by the initial probability measure induced by the preparation φ and the observable A. Correspondingly, the mixed state of the apparatus consists of pointer states Φ_i corresponding to pointer values Z_i with the same weights $p(\varphi, A_i)$.

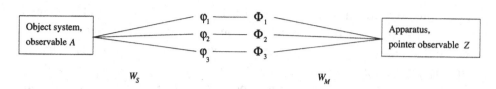

Fig. 1.6 Correlated mixed states W_S and W_M of system S and apparatus M after measurement of observable A with states φ^{A_i} and values A_i. The Φ_i are pointer states that correspond to pointer values Z_i, where A_i and Z_i are connected by the pointer function $A_i = f(Z_i)$.

The two mixed states are strictly correlated, and the corresponding values A_i and Z_i of the observables A and Z are connected by the pointer function $A_i = f(Z_i)$. The slightly simplified diagram in Fig. 1.6 illustrates the situation after the measuring process.

In the realistic interpretation, one further assumes that after the measuring process the pointer possesses some objective value Z_i and the object system has the corresponding A-value $A_i = f(Z_i)$. In other words, the mixed states W_S and W_M are interpreted as meaning that the system and apparatus possess objectively values A_i and Z_i, which are, however, subjectively unknown to the observer. It is obvious that in this situation the observer must merely read off the pointer value in order to complete the measurement. However, this 'ignorance interpretation' of the mixed states W_S and W_M is not justified by quantum theory in any way. For this reason, the many-worlds interpretation dispenses with the objectification assumption and describes the post-measurement situation merely by the entangled pure state $\Psi(S + M)$ or by the two correlated mixed states. Moreover, it will be shown in chapter 4 that an ignorance interpretation of a mixed state is in general even incompatible with the laws of quantum mechanics. It is obvious that this argument strongly supports the restrictive attitude of the many-worlds interpretation.

Without additional assumptions about objectification, the quantum theory of measurement provides a simultaneous description of the complete variety of infinitely many situations of the system such that every value A_i corresponds to some pointer value $Z_i = f^{-1}(A_i)$. The final situation of the measuring process is adequately described by the two correlated mixed states W_S and W_M, one referring to the object system S and one referring to the apparatus M, the 'observer'. The 'measurement' is then nothing else but the correlation between the respective state φ^{A_i} of the system and the 'relative state' Φ_i of the observer-apparatus, which is aware of the object system's state φ^{A_i}. This explains Everett's original terminology 'relative-state interpretation'. The large variety of alternatives that coexist in the final state $\Psi(S + M)$ of the measuring process was later interpreted by some authors as an ensemble of 'really existing worlds' – an idea which has given rise to the name 'many-worlds interpretation' [DeW 70,71], [Whe 57].

The coefficients $p(\varphi, A_i)$ that appear as weights in the mixed states W_S and W_M are also in the present case probabilities in the formal sense. However, in contrast to the realistic and the minimal interpretation, they cannot be interpreted as relative frequencies of the values A_i and Z_i in the mixed states W_S and W_M. The reason is that these mixed states do not express anything other than the state $\Psi(S+M)$ of the whole system $S+M$ after the measuring process and hence do not admit an ignorance interpretation. In other words, there are no objective results A_i and Z_i whose relative frequencies could approach the probability $p(\varphi, A_i)$ for a sufficiently large number of tests.

However, even in this interpretation without objectification the formal probabilities can be given a meaning in the sense of relative frequencies. Let $S^{(N)}$ be an ensemble of N identically prepared systems S_i with states φ and consider this ensemble as a large quantum system. If on each system S_i the observable A is measured, the observer-apparatus will register a sequence of N index numbers $l = \{l_1, l_2, \ldots, l_N\}$ indicating the object values A_{l_k}. These index numbers are then stored in the 'memory' of the observer, where the memory is some registration device of the apparatus. One can then determine the relative frequency of some index value k in the 'memory sequence' l obtained in this way. At this stage of the discussion, one could formulate a new probability reproducibility condition stating that the formal probabilities $p(\varphi, A_i)$ are reproduced in the statistics of the observer's memory sequences. It is not claimed here that the initial probabilities refer to relative frequencies of objective values of the pointer and the object system, respectively. In general, objective values of this kind do not exist. It is merely stated that the formal expressions $p(\varphi, A_i)$ refer to the statistics of a memory sequence obtained by measurements of A on a large number of identically prepared systems.

One might have doubts whether this new requirement is compatible with the laws of quantum mechanics. Moreover, according to the restrictive attitude of the many-worlds interpretation one should even ask for a justification of this postulate by the formalism of quantum mechanics. It was indicated already by Everett [Eve 57] and elaborated in detail by Hartle [Hart 68], Graham [Gra 70], and DeWitt [DeW 71] that the formalism of quantum mechanics does indeed yield the probability interpretation mentioned. These results will be discussed in chapter 3 of the present book, in particular in section 3.5.

Taking account of the results just mentioned, it follows that the many-worlds interpretation is the least restrictive interpretation possible. Indeed, it does not make any additional assumptions going beyond the mere formalism of quantum mechanics. Except for the calibration postulate, which is nothing but a definition of what may be called a measuring apparatus, the many-worlds interpretation can be read off from the formalism. Hence, there is no way to escape the strange consequence of many really existing worlds. This means that quantum mechanics in its present form does not describe the *one* world which we usually have in mind but some reality which is composed of many coexisting distinct worlds.

2

The quantum theory of measurement

2.1 The concept of measurement

2.1(a) Basic requirements

In this chapter we give a brief account of the quantum theory of measurement. As already mentioned in the preceding chapter, the quantum theory of measurement treats the object system, as well as the measuring apparatus, as proper quantum systems. Here we restrict our considerations to a proper quantum mechanical model of the measuring process that makes use of unitary premeasurements. Furthermore, we will be mainly concerned with ordinary discrete observables of the object system that are measured by an apparatus with a pointer observable which is also assumed to be an ordinary discrete observable. These restrictive assumptions are made here and throughout the entire book in order to simplify the problems as much as possible. The remaining open problems of consistency, completeness, self-referentiality, etc. can then be discussed without unnecessary additional complications.

In order to characterize the concept of measurement in quantum mechanics, we formulate some basic requirements that must be fulfilled by any measuring process. In many situations, one can add further postulates, but these additional requirements are not essential for the concept of measurement. The basic requirements are in accordance with the most general interpretation of quantum mechanics, the minimal interpretation, which has already been mentioned in chapter 1. There is a general interplay between interpretation and the quantum theory of measurement, since the postulates that characterize a given interpretation must be compatible with, and capable of being satisfied by, a corresponding model of the measuring process. In particular, the postulates that define the minimal interpretation should be satisfied by the most general model of the measuring process.

The object system S and the apparatus M are assumed to be proper quantum systems with Hilbert spaces $\mathscr{H}_S$ and $\mathscr{H}_M$, respectively. The situation before the measurement is called the *preparation*. Let the object system S be in a pure state $\varphi \in \mathscr{H}_S$ and the pointer of the apparatus M in its neutral state $\Phi \in \mathscr{H}_M$. The observable to be measured is assumed here to be a discrete nondegenerate observable $A = \sum a_i P[\varphi^{a_i}]$ with eigenvalues a_i and eigenstates φ^{a_i}. By $X^A := \{a_i\}$ we denote the value set of A. In general, the preparation state φ of the object system is not an eigenstate of A. Hence, the system S does not possess a value a_i of the observable A prior to the measurement. In chapter 4 it will be shown that it is not possible to assume that a certain value of A pertains objectively to the system in the state φ and that this value is merely subjectively unknown to the observer. For the present problem, this means that the object system $S(\varphi)$ with state φ does not possess an A-value. For this reason, the measuring process cannot be a mere passive observation of objectively decided matters of fact; instead, the measuring process must first prepare the system in such a way that an A-value can be observed in the usual sense. This inevitable active manipulation of the object system will be described here by an interaction Hamiltonian H_{int} of the compound system $S + M$. The initial state $\Psi(S + M)$ of $S + M$ is then changed by a unitary time-dependent operator $U(t) = \exp(-\frac{i}{\hbar}H_{\text{int}}t)$ that acts on $S + M$ within the time interval $0 \le t \le t'$. This interaction part of the measuring process is called *premeasurement*.

In the special case where the preparation φ of the object system S is already an eigenstate of A, i.e., if $\varphi = \varphi^{a_k}$, a measurement of A should lead with certainty to the result $a_k \in X^A$. This is the content of the *calibration postulate* (C_M) of the minimal interpretation, which should be justified by the quantum theory of measurement. The initial state of the compound system is given here by the tensor product $\Psi(S + M) = \varphi^{a_k} \otimes \Phi$ of the pure states $\varphi^{a_k} \in \mathscr{H}_S$ and $\Phi \in \mathscr{H}_M$ of the systems S and M, respectively. The calibration postulate means that the unitary operator $U(t)$ applied to the initial state $\varphi^{a_k} \otimes \Phi$ leads to a state $\Psi'(S + M)$ such that the discrete nondegenerate pointer observable $Z = \sum Z_i P[\Phi_i]$ of the apparatus M assumes the value Z_k that indicates the measurement result $a_k = f(Z_k)$. The values Z_k are the eigenvalues of the pointer observable, and the states $\Phi_i \in \mathscr{H}_M$ are its eigenstates. The pointer function f connects the pointer values with the measurement results and depends on the specific construction of the apparatus. If the unitary operator fulfils these requirements, then the calibration postulate of the minimal interpretation is satisfied.

If in addition the final state $\Psi'(S + M) = U(t)\Psi$ is such that the object system S is in the state φ^{a_k} and the system observable assumes the value

$a_k = f(Z_k)$, then the premeasurement fulfils also the calibration postulate (C_R) of the realistic interpretation. It is obvious that this additional property of the operator $U(t)$, of preserving eigenstates of A, is not necessary for the concept of measurement. Instead, it characterizes the special class of repeatable measurements, which will be discussed later in section 2.4.

If the preparation φ of the object system S is not an eigenstate φ^{a_i} of the observable A that is to be measured, then no prediction can be made about the outcome of an individual measuring process. Hence also no additional requirement can be formulated for the unitary operator U of the premeasurement. However, if after the premeasurement the systems S and M are no longer in interaction, their states are given in general by correlated mixed states W_S' and W_M', respectively. According to the minimal interpretation, the apparatus M in the state W_M' should objectively possess a certain value Z_k of the pointer observable. At the first glance, it is an open question whether this *pointer objectification requirement* (PO) is fulfilled by any unitary premeasurement or whether it leads to a new condition on the unitary operator U. In the case of the realistic interpretation, which requires repeatable measurements, the object system S in the state W_S' should objectively possess the value $a_k = f(Z_k)$ of the system observable A that is indicated by the pointer value Z_k. It is again an open question whether this requirement poses a new condition on unitary and repeatable measurements.

The third postulate of the minimal interpretation, the *probability reproducibility condition* (PR), refers again to the case where the preparation φ of S is not an eigenstate of A. In this case, the pair $\langle \varphi, A \rangle$ consisting of the preparation φ and the observable A induces a probability distribution $p(\varphi, a_i) := |(\varphi^{a_i}, \varphi)|^2$ that fulfils the Kolmogorov axioms of probability. The condition (PR) requires that this initial probability $p(\varphi, a_i)$ be reproduced in the statistics of the pointer values $Z_i = f^{-1}(a_i)$ described by the mixed state W_M' of M after the premeasurement. It turns out that from a formal point of view this condition is fulfilled by any unitary premeasurement and that also the more empirical part of the requirement can be justified without additional assumptions. Similar results hold in the case of the realistic interpretation and repeatable premeasurements for the probability reproducibility condition (SR). The initial probability $p(\varphi, a_i)$ is then reproduced also in the statistics of the system values a_i described by the mixed state W_S' of S after the premeasurement.

The three postulates of calibration, objectification, and reproduction of the initial probability are the basic requirements of the measuring process that constitute the concept of measurement. The apparatus, the pointer observable, the pointer function, and the unitary operator of the interaction must be chosen in such a way that these requirements are fulfilled. It should be emphasized that the concept of measurement formulated here is very weak and minimalistic. Indeed, a measurement result $a_i = f(Z_i)$ that is indicated by a pointer value Z_i does not in general pertain either to the object system in the preparation state or to the system in its state after the measurement. In special cases, the final state of the object may be an eigenstate of A, but this property of repeatable measurements is not part of the general concept of measurement.

2.1(b) Schematic representation

On the basis of the preceding arguments, it is now straightforward to give a brief schematic representation of the measuring process. Considered as a time-dependent process, the measuring process has some similarity to a scattering process of two particles and it consists of *three* steps.

In step I, the *preparation*, the object system S with state φ and the measuring apparatus M with the neutral state Φ, are completely independent. Nevertheless, they will be considered as a compound system $S + M$ with the tensor product state

$$\Psi(S + M) = \varphi(S) \otimes \Phi(M)$$

where $\varphi \in \mathcal{H}_S$ and $\Phi \in \mathcal{H}_M$ are pure states of S and M, respectively.

In step II, the *premeasurement*, the systems S and M are in interaction, which is described here by a unitary operator $U(t) = \exp(-\frac{i}{\hbar}H_{\text{int}}t)$ that acts on the compound state $\Psi(S + M)$ within the time interval $0 \leq t \leq t'$. By H_{int}, we denote the part of the Hamiltonian of $S + M$ that determines the interaction of the two components S and M. Hence, the compound state after the premeasurement is (Fig. 2.1)

$$\Psi'(S + M) = U(t')\varphi \otimes \Phi.$$

The unitary operator U, and thus the Hamiltonian H_{int}, must be chosen such that they are convenient for measuring the system observable $A = \sum a_i P[\varphi^{a_i}]$ with eigenvalues $a_i \in X^A$ and corresponding eigenstates φ^{a_i}. In order to determine the state $\Psi'(S + M)$ further, we apply the *calibration postulate*. According to this postulate, for a system S with the preparation $\varphi = \varphi^{a_i}$ we have

$$U(\varphi^{a_i} \otimes \Phi) = \varphi'_i \otimes \Phi_i. \tag{2.1}$$

Here the states $\Phi_i \in \mathcal{H}_M$ are orthogonal eigenstates of the discrete nondegenerate pointer observable $Z = \sum Z_i P[\Phi_i]$ corresponding to pointer values Z_i. In this almost trivial case, the pointer value Z_i indicates the measurement result $a_i = f(Z_i)$, where the object state φ'_i after the premeasurement need not be an eigenstate of A. This happens only in the case of repeatable measurements, for which one obtains

$$U(\varphi^{a_i} \otimes \Phi) = \varphi^{a_i} \otimes \Phi_i \qquad (2.2)$$

according to the calibration postulate of the realistic interpretation.

If the preparation φ is not an eigenstate of the observable A, then we can make use of the expansion

$$\varphi = \sum_i (\varphi^{a_i}, \varphi)\varphi^{a_i} =: \sum_i c_i \varphi^{a_i} \qquad (2.3)$$

with coefficients $c_i = (\varphi^{a_i}, \varphi)$. Applying the unitary operator $U(t)$ to the compound state $\Psi = \varphi \otimes \Phi$, we obtain

$$U(\varphi \otimes \Phi) = \sum_i c_i\, U(\varphi^{a_i} \otimes \Phi) = \sum_i c_i \varphi'_i \otimes \Phi_i \qquad (2.4)$$

where we have made use of the linearity of U and of eq. (2.1). In the case of repeatable measurements, we use (2.2) and obtain

$$U(\varphi \otimes \Phi) = \sum_i c_i \varphi^{a_i} \otimes \Phi_i. \qquad (2.5)$$

In step III of the measurement, *objectification and reading*, the two systems $S + M$ are again dynamically independent but still correlated. Considered as subsystems of the compound system $S + M$ in the entangled state $\Psi' = \sum c_i \varphi^{a_i} \otimes \Phi_i$ (for repeatable measurements), S and M can be described by the reduced mixed states

$$W'_S = \sum |c_i|^2 P[\varphi^{a_i}], \quad W'_M = \sum |c_i|^2 P[\Phi_i],$$

respectively (Fig. 2.1).

According to the *pointer objectification postulate* (PO), after the measurement the pointer possesses some objective value Z_i indicating the measurement result $a_i = f(Z_i)$. This means that the mixed state W'_M must describe a mixture of states Φ_i such that one of the states Φ_i actually pertains to the apparatus. If we are dealing with repeatable measurements, the strong correlations between S and M after the premeasurement imply the *system objectification postulate* (SO). This means that the mixed state W'_S too describes a mixture of states φ^{a_i} such that the object system S is actually in one of the eigenstates φ^{a_i} of A and possesses the eigenvalue a_i. In this situation, it is no problem to determine the measurement result, a_k, say, that belongs

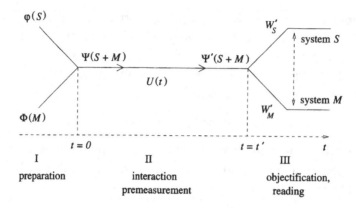

Fig. 2.1 Schematic representation of the measuring process as a time-dependent quantum mechanical two-body problem with systems S and M.

to the final state φ^{a_k}. This is achieved simply by reading the value Z_k of the pointer that is actually realized. Hence, as in classical physics, *reading* concludes the quantum mechanical measuring process (Fig. 2.1).

The measuring process, which has only briefly been sketched in this section and which is illustrated by Fig. 2.1, will be discussed in more detail in the subsequent sections of this chapter.

2.2 Unitary premeasurements

2.2(a) The preparation

The situation before the measurement is characterized by two completely independent proper quantum systems S and M with Hilbert spaces $\mathscr{H}_S$ and $\mathscr{H}_M$, respectively. The object system S is prepared in a pure state $\varphi \in \mathscr{H}_S$ that in general is not an eigenstate of the observable A for which the measurement preparation has just been completed. We will not discuss here the possibilities and intricacies of preparing experimentally the system S in a certain pure state φ. There are many practical difficulties, but there is no fundamental impossibility of preparing a quantum system in a certain pure state. The measuring apparatus is prepared in a pure state $\Phi \in \mathscr{H}_M$ that is 'neutral', i.e., Φ is not an eigenstate of the pointer observable Z of the apparatus. The pointer observable is assumed to be a discrete nondegenerate observable $Z = \sum Z_i P[\Phi_i]$ with orthonormal eigenstates $\Phi_i \in \mathscr{H}_M$. In the case of a macroscopic measuring apparatus, it is of course in practice even more difficult to prepare the system M in a given pure state $\Phi \in \mathscr{H}_M$. However, we will not discuss these technicalities here, since even for pure-state preparation of the system M the measuring process leads to very interesting fundamental problems that have nothing to do with the experimental intricacies just mentioned.

The Hilbert space of the compound system $S + M$ is given by the tensor product $\mathscr{H} = \mathscr{H}_S \otimes \mathscr{H}_M$ of the Hilbert spaces $\mathscr{H}_S$ and $\mathscr{H}_M$. The pure states Ψ of $S + M$ are then elements $\Psi \in \mathscr{H}_S \otimes \mathscr{H}_M$ of the tensor product Hilbert space. If, in particular, the two systems S and M are completely independent and in the states $\varphi \in \mathscr{H}_S$ and $\Phi \in \mathscr{H}_M$, respectively, then the compound system $S + M$ is in the product state $\Psi = \varphi \otimes \Phi$. This is precisely the situation in the first step of the measuring process. Hence, the two systems S and M prepared in the states φ and Φ can be described equivalently by the product state $\varphi \otimes \Phi \in \mathscr{H}$ of the compound system $S + M$.

Some further general remarks on states of compound sytems will be useful for the following considerations and also in subsequent chapters. Let S_1 and S_2 be two systems with Hilbert spaces $\mathscr{H}_1$, $\mathscr{H}_2$ and pure states φ_1, φ_2, respectively. For arbitrary states $\varphi_1 \in \mathscr{H}_1$, $\varphi_2 \in \mathscr{H}_2$ the product state $\varphi_1 \otimes \varphi_2$ is an element of $\mathscr{H}_1 \otimes \mathscr{H}_2$. However, the converse is not true, since an arbitrary state $\Psi(S_1 + S_2) \in \mathscr{H}_1 \otimes \mathscr{H}_2$ cannot be decomposed in general into a product $\varphi_1 \otimes \varphi_2$ of two states $\varphi_1 \in \mathscr{H}_1$ and $\varphi_2 \in \mathscr{H}_2$. If the compound system $S_1 + S_2$ is in an arbitrary pure state $\Psi \in \mathscr{H}_1 \otimes \mathscr{H}_2$, then one can define states W_1 and W_2 of the subsystems S_1 and S_2 in the following way. We call W_1 the state of system S_1 if for an arbitrary observable A_1 of system S_1 the expectation value of A_1 in state W_1 is equal to the expectation value of $A_1 \otimes \mathbb{1}_2$ in the state $\Psi(S_1 + S_2)$, i.e.,

$$(\Psi, A_1 \otimes \mathbb{1}_2 \Psi) = \operatorname{tr}\{A_1 W_1\}. \tag{2.6}$$

Correspondingly, the state W_2 of system S_2 can be defined by the equation

$$(\Psi, \mathbb{1}_1 \otimes A_2 \Psi) = \operatorname{tr}\{A_2 W_2\} \tag{2.7}$$

where A_2 is an arbitrary observable of S_2.

The mathematical elaboration of eqs. (2.6), (2.7) leads to the result that the states W_1 and W_2 are given by the partial traces of $P[\Psi]$ over the degrees of freedom of systems S_2 and S_1, respectively, i.e.,

$$W_1 = \operatorname{tr}_2 P[\Psi], \qquad W_2 = \operatorname{tr}_1 P[\Psi]. \tag{2.8}$$

Here we have used the notation tr_2, say, for the partial trace over a complete set of orthonormal states of $\mathscr{H}_2$. Since the states W_1, W_2 are obtained in (2.8) by reducing the degrees of freedom of $S_1 + S_2$, we also call them *reduced* states. It turns out that the reduced states W_1 and W_2 are not in general pure states – given by projection operators – but mixed states. Considered as a mathematical object, a mixed state W is a self-adjoint positive operator with tr $W = 1$.

There are some exceptions that should be mentioned. A reduced state W_i ($i = 1, 2$) of the (pure) compound state $\Psi(S_1 + S_2)$ is pure if and only if the compound state is of the form $\Psi = \varphi_1 \otimes \varphi_2$. In this case, the reduced state reads $W_i = P[\varphi_i]$. Furthermore, if one of the subsystems, S_1, say, is in a pure state φ_1, then the state of the compound system is $\Psi = \varphi_1 \otimes \varphi_2$ and the other system S_2 is in the pure state φ_2. For details and proofs of these well-known properties of the tensor product, we refer to the literature (e.g. [Jau 68], pp. 179–82).

To calculate explicitly the reduced mixed states W_1 and W_2 given by (2.8), we make use of the following important result (cf. [Neu 32], pp. 231–2): for an arbitrary pure state $\Psi \in \mathscr{H}_1 \otimes \mathscr{H}_2$ of the compound system $S_1 + S_2$ there always exist two complete orthonormal systems $\{\chi_i\}$ and $\{\eta_i\}$, $\chi_i \in \mathscr{H}_1$, $\eta_i \in \mathscr{H}_2$, such that $\Psi(S_1 + S_2)$ can be decomposed as

$$\Psi(S_1 + S_2) = \sum_i c_i \chi_i(S_1) \otimes \eta_i(S_2) \tag{2.9}$$

with coefficients $c_i = (\chi_i \otimes \eta_i, \Psi)$. In contrast to the usual decomposition of a two-body state into a double sum over arbitrary basis systems, the expansion (2.9) contains only a single sum and will be called *biorthogonal*.[†] If the state of the composite system is given in the biorthogonal decomposition (2.9), then one obtains for the partial traces in (2.8)

$$W_1 = \sum |c_i|^2 P[\chi_i], \qquad W_2 = \sum c_i|^2 P[\eta_i]. \tag{2.10}$$

This result shows that the states W_1 and W_2 of the subsystems S_1 and S_2, respectively, are in general mixed states. Considered as mathematical objects W_1 and W_2 are positive self-adjoint operators with tr $W_1 = $ tr $W_2 = 1$. The orthonormal systems $\{\chi_i\}$ and $\{\eta_i\}$ are in general not uniquely determined by the compound state Ψ in (2.9). This is only the case if for all $i, k \in \mathbb{N}$ we have $|c_i| \neq 0$ and $|c_i| \neq |c_k|$ for $i \neq k$. The ambiguity of the orthogonal systems $\{\chi_i\}$ and $\{\eta_i\}$ in the general situation provides serious problems for the interpretation of the mixed states W_1 and W_2 in (2.10) as mixtures of states.

We will come back to this problem in chapter 4.

---

† Some authors call this expansion *normal*, *polar*, or *Schmidt* decomposition. It is based on a mathematical investigation by Schmidt [Schm 07].

2.2(b) Interactions between S and M

In the second step of the measuring process the unitary operator $U(t) = \exp(-\frac{i}{\hbar}H_{\mathrm{int}}t)$ of the premeasurement is applied to the initial state $\Psi = \varphi \otimes \Phi$ at $t = 0$ and acts on the compound system during the time interval $0 \le t \le t'$. Here we use the notation $\Psi(t = 0) = \Psi(S + M) = \varphi \otimes \Phi$ for the state of $S + M$ at $t = 0$ and

$$\Psi(t = t') = \Psi'(S + M) = U(t')\Psi(S + M)$$

for the state at $t = t'$. Since the interaction H_{int} is turned off at $t = t'$, the state of the compound system after the premeasurement is

$$\Psi'(S + M) = U(t')\Psi(S + M), \quad t \ge t'.$$

The compound state is considered here to be time independent. This means that the kinematic time development of the composite system, which is induced by the free two-body Hamiltonian, is not described by the state Ψ'. This can be achieved by using an *interaction representation* such that the time development of $\Psi(t)$ is determined by the interaction Hamiltonian H_{int}, whereas the free part of the Hamiltonian induces a kinematic time dependence of the observables of $S + M$. However, for the majority of practical purposes these subtleties are rather irrelevant, since during the interaction period $0 \le t \le t'$ the energy of the interaction is much larger than the kinetic energy of the system, which can thus be neglected.

According to the calibration postulate, we have

$$U(t')(\varphi^{a_i} \otimes \Phi) = \varphi_i' \otimes \Phi_i \tag{2.11a}$$

in the general case or

$$U(t')(\varphi^{a_i} \otimes \Phi) = \varphi^{a_i} \otimes \Phi_i \tag{2.11b}$$

for repeatable measurements. In the same notation as in eqs. (2.1) and (2.2), the states φ^{a_i} are eigenstates of the measured system observable A, whereas the object states φ_i' after the measurement need not be eigenstates of A. The states $\Phi_i \in \mathcal{H}_M$ are eigenstates of the pointer observable Z. Hence, for an arbitrary preparation $\varphi = \sum c_i \varphi^{a_i}$ of the object system we obtain, on account of the linearity of U, in the general case

$$\Psi' = U(t')(\varphi \otimes \Phi) = \sum_i c_i U(t')(\varphi^{a_i} \otimes \Phi) = \sum_i c_i \varphi_i' \otimes \Phi_i \tag{2.12a}$$

with $\varphi_i' = \varphi^{a_i}$ for repeatable measurements. The coefficients of the expansion are given by $c_i = (\varphi^{a_i}, \varphi)$.

The state Ψ' of $S + M$ after the premeasurement is given here by a single sum

$$\Psi'(S + M) = \sum_i c_i \varphi^{a_i}(S) \otimes \Phi_i(M) \tag{2.12b}$$

over the product states $\varphi^{a_i} \otimes \Phi_i$ (in the case of repeatable measurements), where $\{\varphi^{a_i}\}$ and $\{\Phi_i\}$ are complete orthonormal sets of states in $\mathscr{H}_1$ and $\mathscr{H}_2$, respectively. This is precisely the biorthogonal decomposition of the compound state $\Psi'(S + M)$ mentioned above (eq. (2.9)). It should be noted that the biorthogonal decomposition (2.12b), say, is not obtained here from the given state Ψ' by advanced mathematical techniques (cf. [Neu 32], pp. 231–2 or [Jau 68], pp. 179–82). Instead, the representation (2.12b) evolves by itself if one starts from the calibration condition (2.11b).

On the basis of the compound state Ψ' in the decompositions (2.12a) or (2.12b), the reduced mixed states W'_S and W'_M of the subsystems S and M, respectively, can easily be determined. If we restrict our consideration again to repeatable measurements, we obtain by means of formula (2.10)

$$W'_S = \sum |c_i|^2 P[\varphi^{a_i}], \qquad W'_M = \sum |c_i|^2 P[\Phi_i] \tag{2.13}$$

where the coefficients are given by $|c_i|^2 = |(\varphi^{a_i}, \varphi)|^2$. The expressions W'_S and W'_M describe the states of systems S and M, respectively, after the premeasurement and are uniquely determined by the state $\Psi'(S + M)$ of the compound system. However, it is obvious that the decomposition of W'_S and of W'_M into orthogonal components given by (2.13) is unique only if all the coefficients $|c_i|^2$ are different and nonvanishing. For this reason, in the general case there are difficulties in an attempt to interpret the mixed states W'_S and W'_M as mixtures of the pure states φ^{a_i} and Φ_i, respectively, with weights $|c_i|^2$.

For a given observable A, one needs some unitary operator U_A that is convenient for measuring this observable for arbitrary preparations φ of S. It is an important question whether an operator U_A of this kind always exists. The first attempt to answer this question in the affirmative was made by von Neumann ([Neu 32], p. 235). For a given observable A, von Neumann constructed a unitary operator U_A and demonstrated in this way the existence of a measuring interaction for arbitrary observables. However, the operator U_A that was constructed by von Neumann is rather formal and artificial and far from any intuitive and realistic expression for an interaction operator.

A more explicit example for the operator U_A and even for the interaction Hamiltonian H_{int} can be given in the following way. Let P^Z be the observable of the apparatus M that is canonically conjugate to the pointer observable

Z. Hence, Z and P^Z fulfil the canonical commutation relation

$$[Z, P^Z] = i\hbar.$$

A model for the unitary operator U_A is then given by

$$U_A = e^{-\frac{i}{\hbar} H_{\text{int}} \Delta t} = e^{i\lambda(A \otimes P^Z)}. \tag{2.14}$$

The parameter λ is proportional to the time interval Δt and to a coupling constant that indicates the strength of the interaction. For the Hamiltonian, we obtain

$$H_{\text{int}} = -\frac{\lambda\hbar}{\Delta t}(A \otimes P^Z).$$

It remains to be shown that the operator U_A fulfils the calibration condition. If we apply U_A to the state $\Psi(S + M) = \varphi^{a_i} \otimes \Phi$, we obtain

$$e^{i\lambda(A \otimes P^Z)}\varphi^{a_i} \otimes \Phi = \varphi^{a_i} \otimes e^{i\lambda a_i P^Z}\Phi. \tag{2.15}$$

The operator $e^{i\lambda a_k P^Z}$ shifts the pointer observable Z according to

$$e^{-i\lambda a_k P^Z} Z\, e^{+i\lambda a_k P^Z} = Z - \lambda a_k \mathbb{1}.$$

If we apply $e^{i\lambda a_k P^Z}$ to the neutral pointer state Φ with value Z_0, i.e., with $Z\Phi = Z_0\Phi$, we obtain the state

$$\Phi^{(k)} := e^{i\lambda a_k P^Z}\Phi$$

which fulfils the equation $Z\Phi^{(k)} = (Z_0 - \lambda a_k)\Phi^{(k)}$ and which is thus an eigenstate of Z with eigenvalue $Z_0 - \lambda a_k$. For the convenient choice of pointer function $f(Z_k) := \frac{Z_0 - Z_k}{\lambda} = a_k$, it follows that

$$Z\Phi^{(k)} = Z_k\Phi^{(k)}$$

and hence $\Phi^{(k)} = \Phi_k$. Inserting this result into (2.15), we obtain

$$e^{i\lambda(A \otimes P^Z)}\varphi^{a_i} \otimes \Phi = \varphi^{a_i} \otimes \Phi_i \tag{2.16}$$

in accordance with the calibration postulate. This means that the unitary operator U_A in (2.14) fulfils the requirements of a unitary premeasurement of A. Realistic examples for operators of the form (2.14) can be found in the literature (e.g. [BGL 95], chapter 7).

2.3 Classification of premeasurements

In this section, we characterize some classes of premeasurements by properties of the corresponding unitary operators. Here we will not try to construct explicit expressions for the unitary operators like that for the standard model (2.15). Instead, we describe the operators U by their effect on states $\varphi^{a_i} \otimes \Phi$ that are used in the calibration postulate. We will investigate again measurements of a discrete observable A with eigenvalues a_i, but we will not restrict our considerations to the case of nondegenerate observables. Furthermore, we will also briefly discuss the measurement of continuous observables and the special problems that arise in this case. More general types of premeasurements can be found in [BLM 91].

2.3(a) Discrete observables

Type 1. Standard measurements. Let A be a discrete and nondegenerate observable $A = \sum_i a_i P[\varphi^{a_i}]$ with eigenstates φ^{a_i}, eigenvalues a_i, and a value set $\{a_i\} = X^A$. The simplest kind of unitary premeasurement is given by a unitary operator U such that for all states φ^{a_i} the relation

$$U(\varphi^{a_i} \otimes \Phi) = \varphi^{a_i} \otimes \Phi_i \tag{2.17}$$

holds. The initial states φ^{a_i} of the object system S are not altered at all by this kind of measurement. The initial (neutral) state Φ of the pointer is transformed into the eigenstate Φ_i of the pointer that corresponds to the pointer value Z_i and that indicates the measurement result $a_i = f(Z_i)$. We add that, for an arbitrary preparation φ, the state of $S + M$ after the premeasurement reads

$$U(\varphi \otimes \Phi) = \sum (\varphi^{a_i}, \varphi) \varphi^{a_i} \otimes \Phi_i$$

and the corresponding mixed state $W_S'(\varphi, A)$ of the object system is given by

$$W_S'(\varphi, A) = \sum_i p^A(\varphi, a_i) P[\varphi^{a_i}],$$

where $p^A(\varphi, a_i) = |(\varphi^{a_i}, \varphi)|^2$ is the probability for obtaining A_i as the result of the A-measurement. This very simple standard type has been discussed already as an illustration of the measuring process.

Type 2. Lüders measurements. The next generalization consists of assuming that A has a degenerate spectrum and can be expressed as

$$A = \sum a_i P(a_i) = \sum a_i \sum_{j=1}^{n_i} P[\varphi_j^{a_i}]. \tag{2.18}$$

Here $P(a_i)$ is the projection operator that projects onto the subspace (closed linear manifold) $M(a_i) := P(a_i)\mathcal{H}_S$, which belongs to the eigenvalue a_i. The dimension of this subspace is given by $\dim M(a_i) = \mathrm{tr}\{P(a_i)\} = n_i \geq 1$. In the subspace $M(a_i)$, one can introduce an orthonormal basis $\varphi_j^{a_i}$ ($j = 1, \ldots, n_i$) such that $P(a_i) = \sum_{j=1}^{n_i} P[\varphi_j^{a_i}]$. The eigenvalue equation that corresponds to the decomposition (2.18) then reads

$$A\varphi_j^{a_i} = a_i \varphi_j^{a_i}, \qquad \varphi_j^{a_i} \in M(a_i).$$

In analogy to (2.17) for the unitary premeasurement, we write in this case

$$U(\varphi_j^{a_i} \otimes \Phi) = \varphi_j^{a_i} \otimes \Phi_i. \tag{2.19}$$

With the expansion $\varphi = \sum_i^\infty \sum_j^{n_i} c_{ij} \varphi_j^{a_i}$ we obtain

$$U(\varphi \otimes \Phi) = \sum_i^\infty \sum_j^{n_i} c_{ij} \varphi_j^{a_i} \otimes \Phi_i, \tag{2.20}$$

where $c_{ij} = (\varphi_j^{a_i}, \varphi)$. Equation (2.20) can be brought into a biorthogonal form by introducing the normalized states

$$\varphi^{(a_i)} := \frac{1}{N_i} P(a_i)\varphi = \frac{1}{N_i} \sum_j^{n_i} c_{ij} \varphi_j^{a_i}$$

with normalization constants $N_i^2 = (\varphi, P(a_i)\varphi) = \sum_j^{n_i} |c_{ij}|^2$. Obviously, a state $\varphi^{(a_i)}$ is the normalized projection of φ onto the subspace $M(a_i)$, and N_i^2 is the probability $p(\varphi, a_i)$ for measuring the value a_i. The state $\Psi'(S + M)$ thus reads

$$U(\varphi \otimes \Phi) = \sum_i N_i \varphi^{(a_i)} \otimes \Phi_i, \tag{2.21}$$

where the states $\varphi^{(a_i)} \in M(a_i)$ are orthogonal but not complete. The measurement (2.21) is called a *Lüders measurement* and the states $\varphi^{(a_i)}$ *Lüders states*. The mixed state W_S' of S after a Lüders premeasurement of A reads $W_{S,L}'(\varphi, A) = \sum p(\varphi, a_i) P[\varphi^{(a_i)}]$ and is called a *Lüders mixture*. It is obvious that the final state of S after reading is given by one of the states $\varphi^{(a_i)}$.

Type 3. Strong value-correlation measurements. The Lüders measurement (2.19), (2.21) of a degenerate observable A as shown in (2.18) can be generalized if we replace the unitary premeasurement (2.19) by

$$U(\varphi_j^{a_i} \otimes \Phi) = \psi_j^{a_i} \otimes \Phi_i. \tag{2.22}$$

Here the states $\psi_j^{a_i} \in \mathcal{H}_S$ are orthonormal with respect to both indices, i.e., $(\psi_j^{a_i}, \psi_l^{a_k}) = \delta_{ik}\delta_{jl}$, and are eigenstates of A such that

$$A\psi_j^{a_i} = a_i \psi_j^{a_i}, \tag{2.23}$$

but the $\psi_j^{a_i}$ need not be identical to the $\varphi_j^{a_i}$. For an arbitrary preparation $\varphi = \sum c_{ij}\varphi_j^{a_i}$ we obtain

$$U(\varphi \otimes \Phi) = \sum_{i,j} c_{ij}\psi_j^{a_i} \otimes \Phi_i = \sum N_i \gamma^{a_i} \otimes \Phi_i \qquad (2.24)$$

with normalized states $\gamma^{a_i} = N_i^{-1} \sum c_{ij}\psi_j^{a_i}$ and coefficients $N_i^2 = \sum |c_{ij}|^2 = p(\varphi, a_i)$. The new states $\gamma^{a_i} \in M(a_i)$ are orthonormal eigenstates of A, $A\gamma^{a_i} = a_i\gamma^{a_i}$, but in general they are not identical to the Lüders states $\varphi^{(a_i)}$, which are defined as the normalized projections of φ onto the subspaces $M(a_i)$. However, since (2.24) is again a biorthogonal expansion, the mixed states after the premeasurement read

$$W'_{S,V} = \sum p(\varphi, a_i) P[\gamma^{a_i}], \qquad W'_{M,V} = \sum p(\varphi, a_i) P[\Phi_i] \qquad (2.25)$$

and there is a strong correlation between the values a_i and Z_i. For this reason, type-3 measurements are called strong value-correlation measurements.

Type 4. Strong state-correlation measurements. A generalization of the measurement (2.22) is given by

$$U(\varphi_j^{a_i} \otimes \Phi) = \psi_{ij} \otimes \Phi_i \qquad (2.26)$$

with orthonormal states ψ_{ij} that need not be eigenstates of A. For an arbitrary preparation φ, we obtain again the biorthogonal expansion

$$U(\varphi \otimes \Phi) = \sum N_i \gamma_i \otimes \Phi_i$$

with orthonormal states $\gamma_i := N_i^{-1} \sum_j c_{ij}\psi_{ij}$, $N_i^2 = p(\varphi, a_i)$, which are in general not eigenstates of A. There is still a strong state-correlation between the states Φ_i and γ_i, but the states γ_i have no direct physical meaning. The mixed states after this measurement,

$$W'_{S,S} = \sum p(\varphi, a_i) P[\gamma_i], \qquad W'_{M,S} = \sum p(\varphi, a_i) P[\Phi_i],$$

show that the pointer states Φ_i and the values Z_i have probabilities $p(\varphi, a_i)$ but the states γ_i of the object system have no direct connections to the values $a_i = f(Z_i)$ that are obtained from the pointer values Z_i. The strong state-correlation measurement is the simplest example of a unitary premeasurement that fulfils the postulates of the minimal interpretation, but that does not provide any useful information about the object system after the measurement.

It is obvious that one could proceed to even more general types of premeasurements with nonorthogonal states γ_i. We will not go into details here and refer instead to the literature [BLM 91]. The essential point is that there are unitary premeasurements in accordance with the minimal interpretation for which the result a_i has nothing to do with the state of the object system after the measurement. Since the measured value $a_i \in X^A$ in general does not pertain to S before the measurement, we are confronted here with the strange situation that the measurement result $a_i = f(Z_i)$ does not provide any information about the A-value of the object system before or after the measurement.

2.3(b) Continuous observables

If the observable A has a continuous spectrum, we write for the spectral decomposition

$$A = \int_{\mathbb{R}} a \, dP^A(a), \qquad a \in \mathbb{R}.$$

The value set of A is given here by $X^A = \mathbb{R}$. By means of Borel subsets $X \in \mathscr{B}(\mathbb{R})$, we can express the probabilities

$$p^A(\varphi, X) = (\varphi, P^A(X)\varphi)$$

for obtaining a measured result in X.

For a discussion of the measuring process, we discretize the observable A by a fixed partition (X_i) of the value set $X^A = \mathbb{R}$ of the system into mutually disjoint Borel subsets $X_i \in \mathscr{B}(\mathbb{R})$ such that $X^A = \cup X_i$. This discretization corresponds to a countable decomposition of the value set $X^A = \mathbb{R}$. Clearly, there are uncountably many partitions of this kind, and they will be labelled here by some index value $\lambda \in \mathbb{R}$. Any such partition $(X_i)_\lambda$ of the value set defines a discrete A-observable $A^\lambda = \sum_i X_i^\lambda P^A(X_i^\lambda)$ and a reading scale R_λ that defines a discretized version $Z^\lambda = \sum Z_i^\lambda P^Z(Z_i^\lambda)$ of the pointer observable Z, where the values $Z_i^\lambda = f^{-1}(X_i^\lambda)$ are given by a convenient pointer function.

For any fixed partition $(X_i)_\lambda$ of the value set X^A and the corresponding discrete observables A^λ and Z^λ, we can apply the full variety of possible types of premeasurements described in the previous subsection. In this way, we finally obtain some pointer value Z_i^λ that indicates a certain Borel subset $X_i^\lambda \in \mathscr{B}(\mathbb{R})$ as the measurement result. In the general case, nothing is known about the state and the A-value of the object system S. The more general remarks about the meaning of a measurement result for a discrete observable also apply here – *mutatis mutandis* – to continuous observables.

2.3(c) General remarks

The various types of unitary premeasurements discussed here have shown that a certain measurement can be characterized by a number of necessary ingredients. In all cases, we need an apparatus Hilbert space $\mathscr{H}_M$, a pointer observable Z, the initial pointer state Φ, a unitary operator U_A that is convenient for measuring A, and a pointer function f that correlates the pointer values Z_i with the measurement results $a_i \in X^A$. The collection $\langle \mathscr{H}_M, Z, \Phi, U_A, f \rangle$ of components, which may be extended by other elements, will be called here a *measurement scheme*. A measurement scheme is the most general description of the measuring process that can be applied to object systems with arbitrary preparations.

In order to characterize adequately the various types of premeasurements, we introduce two concepts that are frequently used for the classification of measurements: *ideality* and *repeatability*. The second of these concepts, repeatability, has already been used in an intuitive sense. In the following, we will give more-formal definitions of both these concepts.

1. A measurement is called ideal if it alters the measured object system only to the extent that is necessary for the measurement. According to the discussion in section 2.1, a quantum mechanical measuring process is not a mere passive observation of objectively decided facts but an active manipulation of the system. Hence, quantum measurements cannot be ideal in the sense that the object is not altered at all. For this reason, ideality means here that all properties that pertain to the system in the preparation state and are compatible with the measured observable pertain to the system also after the premeasurement. For the initial state (the preparation) of S we write φ and for the mixed state of S after the premeasurement of A we write $W_S'(\varphi, X^A)$, indicating that the result X is an arbitrary element of X^A.

A measurement of A is then ideal if for any state φ and for a property B that is compatible with A, i.e., if $[A, B] = 0$,

$$p(\varphi, B) = 1 \quad \text{implies} \quad p(W_S'(\varphi, X^A), B) = 1. \tag{2.27a}$$

Applied to the observable A itself, this means that for any state φ and all values a_i

$$p(\varphi, a_i) = 1 \quad \text{implies} \quad p(W_S'(\varphi, X^A), a_i) = 1. \tag{2.27b}$$

Note that in a slight modification of our previous terminology we have used here the mixed state $W_S'(\varphi, X^A)$ as an argument in the probability function p.

2. A measurement of the observable A is called repeatable if the repetition of the measurement does not change the result. If the preparation φ is not an eigenstate of the measured observable A, then the result $a_i \in X^A$ of a quantum mechanical measurement cannot be predicted with certainty but only with some probabiliy $p(\varphi, a_i)$, $a_i \in X^A$. Hence, a formal definition of the concept of repeatability must make use of probabilities. If we denote by $W_S'(\varphi, a_i)$ the normalized component of $W_S'(\varphi, X^A)$ that corresponds to the measured result $a_i \in X^A$, then we arrive at the following definition: a measurement of A is called *repeatable* if for any state φ and for all values $a_i \in X^A$

$$p(W_S'(\varphi, a_i), a_i) = 1. \tag{2.28}$$

In other words, the probability $p(W_S'(\varphi, a_i), a_i)$ for obtaining the value a_i of A as a result of the repeated A-measurement is equal to 1 if the result of the previous measurement was a_i.

The concepts of *ideality* and *repeatability* can easily be applied to the various types of premeasurements discussed. Without explicit calculation, we mention here the following results:

Measurements of type 1 are ideal and repeatable;
Measurements of type 2 are ideal and repeatable;
Measurements of type 3 are not ideal but are repeatable;
Measurements of type 4 are not ideal and not repeatable.

Finally, we mention a very important result about measurements of continuous observables. According to a theorem by Ozawa [Oza 84], an observable that admits a repeatable measurement is a discrete observable. In other words, for a continuous observable like position or momentum there does not exist a repeatable measurement. It is a further consequence of this result that for a continuous observable there does not exist an ideal measurement in the sense defined above.

2.4 Separation of object and apparatus

2.4(a) Mixed states versus mixtures of states

After the premeasurement, the object system S and the apparatus M are – considered as isolated systems – in the reduced mixed states W_S' and W_M'. In general, these states can be obtained from the compound state Ψ' by partial traces, according to (2.8). In the special case of repeatable type-1 measurements, one gets the expressions (2.13), which will also be considered here. The states W_S' and W_M' are self-adjoint operators, the spectral decompositions of which are given by (2.13). The spectral decompositions and thus the states φ^{a_i} and Φ_i are unique except when the operators W_S' and W_M' have a degenerate spectrum. Since this is not the case here, the spectral decompositions considered are in fact unique.

In step III of the measuring process, objectification and reading (subsection 2.1(b)), the mixed states

$$W_S' = \sum |c_i|^2 P[\varphi^{a_i}], \qquad W_M' = \sum |c_i|^2 P[\Phi_i]$$

are interpreted as *mixtures of states*

$$\Gamma_S(|c_i|^2, \varphi^{a_i}), \qquad \Gamma_M(|c_i|^2, \Phi_i)$$

with weights $|c_i|^2$. These mixtures of states, also called *Gemenge*, have the following meaning. If system S, say, is described by a mixture of states $\Gamma(|c_i|^2, \varphi^{a_i})$, then this system is in one of the states φ^{a_i} with probability $|c_i|^2$. This means that one of the values a_i pertains objectively to the system, but this value is unknown to the observer, who knows only its probability.

From a formal point of view, this Gemenge interpretation of the mixed states W_S' and W_M' is exposed to the objection that the decomposition of W_S', say, is not unique. If W_S' is not degenerate, then the spectral decomposition into orthogonal states φ^{a_i} is unique. However, even in the case of a nondegenerate operator W_S' there are infinitely many decompositions into nonorthogonal states (cf. [BeCa 81] pp. 9–11). Consequently, the interpretation of the mixed state $W_S' = \sum |c_i|^2 P[\varphi^{a_i}]$ as a mixture of states φ^{a_i} with weights $|c_i|^2$ is not a mathematical corollary that follows from the state operator W_S' itself. The postulates that must be added here are the pointer and system objectification requirements (PO) and (SO), respectively.

We will not discuss here the question whether the objectification requirements can be fulfilled in some way. This problem will be discussed extensively in chapters 4 and 5. For the present considerations, it is sufficient to emphasize the following point: if objectification has been achieved and the Gemenge interpretation of the mixed state is justified, then there is no further problem in reading and registering the final result. The Gemenge $\Gamma(|c_i|^2, \varphi^{a_i})$ is a classical mixture of states with probabilities $p_i = |c_i|^2$. For this reason, the final measurement result a_k, say, can be read off from the pointer without any further alteration of the system and the apparatus. The last step of the measuring process, reading the result, merely reduces the subjective ignorance of the observer.

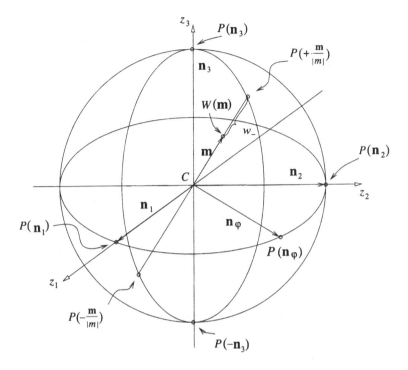

Fig. 2.2 Poincaré sphere $\mathscr{P}$ with centre C, orthogonal coordinates z_1, z_2, z_3, and unit vectors $\mathbf{n}_1, \mathbf{n}_2, \mathbf{n}_3$. Any surface point given by a unit vector $\mathbf{n}_\varphi$ corresponds to a pure state $P(\mathbf{n}_\varphi)$, while points in the interior of $\mathscr{P}$ and vectors $\mathbf{m}$ with $|\mathbf{m}| < 1$ correspond to mixed states $W(\mathbf{m})$.

2.4(b) Illustration: The Poincaré sphere

The measuring process in its standard version (type 1) can be illustrated graphically if one restricts consideration to discrete nondegenerate observables with only two eigenvalues and two eigenstates. The full measuring process – as far as it concerns the object system – can then be described in a two-dimensional Hilbert space $\mathscr{H}_2$. The graphical representation of the Hilbert space $\mathscr{H}_2$ will be considered here in somewhat more detail, since this way of illustration will be used not only for the measuring process but also on several other occasions in subsequent chapters.

Pure and mixed-state operators of a two-dimensional Hilbert space $\mathscr{H}_2$ can be represented by means of a three-dimensional unit sphere, the *Poincaré* sphere $\mathscr{P}$ (Fig. 2.2). Using the Pauli operators $\{\mathbb{1}, \sigma_1, \sigma_2, \sigma_3\}$ in $\mathscr{H}_2$ with properties

$$\sigma_j\sigma_k - \sigma_k\sigma_j = 2\varepsilon_{jkl}\sigma_l, \qquad \sigma_k\sigma_k + \sigma_k\sigma_j = 2\delta_{jk}\mathbb{1}, \qquad (2.29)$$

we can represent projection operators $P(\mathbf{x})$ by

$$P(\mathbf{x}) = \tfrac{1}{2}(x_i\sigma_i + \mathbb{1}), \quad x_i \in \mathbb{R}, \quad \sum x_i^2 = 1 \tag{2.30}$$

and mixed states by

$$W(\mathbf{y}) = \tfrac{1}{2}(y_i\sigma_i + \mathbb{1}), \quad y_i \in \mathbb{R}, \quad \sum y_i^2 < 1. \tag{2.31}$$

From (2.29) it follows that $\sigma_i^2 = \mathbb{1}$, and thus ± 1 for the eigenvalues of the Pauli operators.

Geometrically, projection operators $P(\mathbf{x})$ are given by unit vectors $\mathbf{x} = (x_1, x_2, x_3)$, or points on the surface of $\mathscr{P}$, and mixed states by vectors $\mathbf{y}$ with $|\mathbf{y}| < 1$, or points in the interior of $\mathscr{P}$. The projection operators $P(\mathbf{x})$ and $P(-\mathbf{x}) = \mathbb{1} - P(\mathbf{x})$ project onto orthogonal states and correspond to diametrically opposite points of the Poincaré sphere $\mathscr{P}$.

The unit vectors in the directions of the three orthogonal coordinate axes z_i will be denoted by $\mathbf{n}_1, \mathbf{n}_2, \mathbf{n}_3$. Hence, the vector $\mathbf{n}_3 = (0, 0, 1)$, say, describes the projection operator

$$P(\mathbf{n}_3) = \tfrac{1}{2}(\mathbf{n}_3 \cdot \boldsymbol{\sigma} + \mathbb{1}) = \tfrac{1}{2}(\sigma_3 + \mathbb{1}), \tag{2.32}$$

which corresponds to the north pole of the Poincaré sphere. Similarly, the south pole is given by $P(-\mathbf{n}_3) = \mathbb{1} - P(\mathbf{n}_3)$. The projection operator $P(\mathbf{n}_1)$ given by the vector $\mathbf{n}_1$ (orthogonal to $\mathbf{n}_3$) is

$$P(\mathbf{n}_1) = \tfrac{1}{2}(\sigma_1 + \mathbb{1}). \tag{2.33}$$

The one-dimensional projection operators $P(\mathbf{n})$ may be considered either as pure states or as observables with eigenvalues 0 and 1. The projection operators $P(\mathbf{n})$ and $P(\mathbf{n}')$ with orthogonal vectors $\mathbf{n}$ and $\mathbf{n}'$ ($\mathbf{n} \cdot \mathbf{n}' = 0$) correspond to complementary observables. Hence, $P(\mathbf{n}_1)$ and $P(\mathbf{n}_3)$ are complementary. A pure state $\varphi \in \mathscr{H}_2$ that is given by a unit vector $\mathbf{n}_\varphi$ and represented by the projector $P(\mathbf{n}_\varphi) = P[\varphi]$ is an eigenstate of the observable $P(\mathbf{n})$ if $\mathbf{n} \cdot \mathbf{n}_\varphi = 1$ (eigenvalue 1) or if $\mathbf{n} \cdot \mathbf{n}_\varphi = -1$ (eigenvalue 0). Hence, the vectors $\mathbf{n}_\varphi$ of eigenstates are either parallel or antiparallel to the vector $\mathbf{n}$ of the observable considered.

Let φ_1 and φ_2 be two arbitrary pure states of $\mathscr{H}_2$ given by the unit vectors $\mathbf{n}_{\varphi_1}$ and $\mathbf{n}_{\varphi_2}$ in $\mathscr{P}$. The scalar product (φ_1, φ_2) in $\mathscr{H}_2$ is then related to the vectors $\mathbf{n}_{\varphi_1}$ and $\mathbf{n}_{\varphi_2}$ in the Poincaré sphere by the formula

$$|(\varphi_1, \varphi_2)|^2 = \tfrac{1}{2}(1 + \mathbf{n}_{\varphi_1} \cdot \mathbf{n}_{\varphi_2}). \tag{2.34}$$

In particular, if $P(\mathbf{n})$ is an observable and $P(\mathbf{n}_\varphi)$ a state, one obtains for the probabilities $p(\varphi, 1)$ and $p(\varphi, 0)$ of the eigenvalues 1 and 0 with respect to φ

$$p(\varphi, 1) = \tfrac{1}{2}(1 + \mathbf{n} \cdot \mathbf{n}_\varphi) \tag{2.35a}$$

and

$$p(\varphi, 0) = \tfrac{1}{2}(1 - \mathbf{n} \cdot \mathbf{n}_\varphi). \tag{2.35b}$$

Mixed states $W(\mathbf{m})$ are given according to (2.31) by vectors $\mathbf{m}$ with $|\mathbf{m}| < 1$, i.e., points in the interior of the Poincaré sphere. Geometrically, the spectral decomposition of $W(\mathbf{m})$ can be obtained by drawing the diameter through the point $W(\mathbf{m})$. Formally, the spectral decomposition follows from (2.31) and is

$$W(\mathbf{m}) = w_+ P\left(+\frac{\mathbf{m}}{|\mathbf{m}|}\right) + w_- P\left(-\frac{\mathbf{m}}{|\mathbf{m}|}\right) \tag{2.36}$$

with $w_+ + w_- = 1$. The eigenvalues w_+ and w_- are given by

$$w_\pm = \tfrac{1}{2}(1 \pm |\mathbf{m}|) \tag{2.37}$$

and can be interpreted as the probabilities of the (pure) eigenstates of $W(\mathbf{m})$ that are given by the antiparallel unit vectors $\pm\frac{\mathbf{m}}{|\mathbf{m}|}$ and the projection operators

$$P\left(\pm\frac{\mathbf{m}}{|\mathbf{m}|}\right) = \frac{1}{2}\left(\pm\frac{m_i}{|\mathbf{m}|}\sigma_i + 1\right).$$

In the Poincaré sphere, the eigenvalues $w_\pm$ of $W(\mathbf{m})$ have an immediate geometrical meaning. On the diameter through the point W, the distances $d_\pm$ between W and the points $P\left(\pm\frac{\mathbf{m}}{|\mathbf{m}|}\right)$ on the surface of $\mathscr{P}$ are related to the eigenvalues by $d_\pm = 2w_\mp$.

Within the framework of the Poincaré sphere, the measuring process can be easily illustrated. Let

$$A := P(\mathbf{n}_1) = \tfrac{1}{2}(\sigma_1 + 1)$$

be the measured observable and

$$W := P(\varphi) = \tfrac{1}{2}(\varphi_1 \sigma_1 + \varphi_2 \sigma_2 + 1)$$

the preparation of the system. Clearly, the vector $\mathbf{n}_\varphi$ lies in the equatorial plane, and the angle δ between the vectors $\mathbf{n}_1$ and $\mathbf{n}_\varphi$ is given by $\cos\delta = \mathbf{n}_1 \cdot \mathbf{n}_\varphi = \varphi_1$. The unitary premeasurement of A leads to the mixed state

$$W'(\varphi, A) = p(\varphi, 1)P(+\mathbf{n}_1) + p(\varphi, 0)P(-\mathbf{n}_1),$$

where $p(\varphi, 1)$ and $p(\varphi, 0)$ are the probabilities of the eigenvalues 1 and 0 of the observable A, and $P(+\mathbf{n}_1)$, $P(-\mathbf{n}_1)$ are the corresponding eigenstates. According to (2.35a,b) the probabilities are

$$p(\varphi, 1) = \tfrac{1}{2}(1 + \mathbf{n}_1 \cdot \mathbf{n}_\varphi) = \tfrac{1}{2}(1 + \cos\delta)$$
$$p(\varphi, 0) = \tfrac{1}{2}(1 - \mathbf{n}_1 \cdot \mathbf{n}_\varphi) = \tfrac{1}{2}(1 - \cos\delta).$$

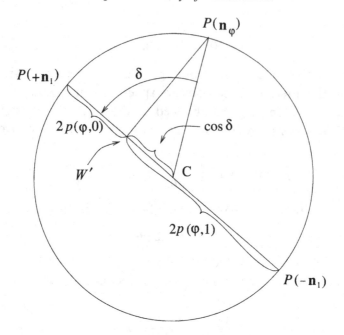

Fig. 2.3 Illustration of the measuring process in the Poincaré sphere: preparation $P(\mathbf{n}_\varphi)$, measured observable $P(\mathbf{n}_1)$, mixed state W', and probabilities $p(\varphi,0)$, $p(\varphi,1)$.

The mixed state $W'(\varphi, A)$ is given by a point in the interior of the Poincaré sphere, or by the vector $\mathbf{m} = (\mathbf{n}_1 \cdot \mathbf{n}_\varphi)\,\mathbf{n}_1 = \cos\delta\,\mathbf{n}_1$, i.e.,

$$W'(\mathbf{m}) = \tfrac{1}{2}(m_1\sigma_1 + \mathbb{1}).$$

Geometrically, W' can be obtained by orthogonal projection of the point $P(\mathbf{n}_\varphi)$ onto the diameter in the direction $\mathbf{n}_1$ (Fig. 2.3). The distance between the centre C of the sphere and the point W' is given by the length $|\mathbf{m}| = \cos\delta$ of the vector $\mathbf{m}$, and the distances d_+ and d_- between W' and the points $P(+\mathbf{n}_1)$ and $P(-\mathbf{n}_1)$ are, respectively,

$$d_+ = 1 - \cos\delta = 2p(\varphi,0), \qquad d_- = 1 + \cos\delta = 2p(\varphi,1)$$

and illustrate geometrically the measurement probabilities.

In the last step of the measuring process, which was called *reading*, the observer recognizes that the eigenvalue 1, say, and the corresponding final object state $P(+\mathbf{n}_1)$, are realized.

3

The probability interpretation

3.1 Historical remarks

The quantum mechanical formalism discovered by Heisenberg [Heis 25] and Schrödinger [Schrö 26] in 1925 was first interpreted in a statistical sense by Born [Born 26]. The formal expressions $p(\varphi, a_i) = |(\varphi, \varphi^{a_i})|^2$, $i \in \mathbb{N}$, were interpreted as the probabilities that a quantum system S with preparation φ possesses the value a_i that belongs to the state φ^{a_i}. This original Born interpretation, which was formulated for scattering processes, was, however, not tenable in the general case. The probabilities must not be related to the system S in state φ, since in the preparation φ the value a_i of an observable A is in general not subjectively unknown but objectively undecided. Instead, one has to interpret the formal expressions $p(\varphi, a_i)$ as the probabilities of finding the value a_i after measurement of the observable A of the system S with preparation φ. In this improved version, the statistical or Born interpretation is used in the present-day literature.

On the one hand, the statistical (Born) interpretation of quantum mechanics is usually taken for granted, and the formalism of quantum mechanics is considered as a theory that provides statistical predictions referring to a sufficiently large ensemble of identically prepared systems $S(\varphi)$ after the measurement of the observable in question. On the other hand, the meaning of the same formal terms $p(\varphi, a_i)$ for an individual system is highly problematic. Heisenberg [Heis 59] tried to understand the meaning of the expressions $p(\varphi, a_i)$ for a single system by means of the Aristotelian concept of 'potentia'. Popper introduced the new concept of 'propensity' for the same reason. However, since neither of these attempts to understand probability on the individual level gives rise to any observable prediction, in the current literature the individualistic interpretation of quantum mechanics is not considered as an alternative that should be taken seriously – with the notable exception of the work of Shimony [Shim 89].

In contrast to this well-known situation, recent results on quantum probability seem to be more in favour of the opposite way of reasoning. Quantum mechanics is then primarily a theory of individual systems, and the statistical predictions of the same theory merely result from the behaviour of a large number of individual systems. If this way of reasoning proves to be correct, the statistical (Born) interpretation of quantum mechanics would turn out to be merely a preliminary way of speaking about a domain of reality that can in principle also be understood on the individualistic level. In the present chapter this way of reasoning will be elaborated in more detail.

The formal results of this chapter are valid for any sharp quantum mechanical observable and for any unitary measurement. However, for the sake of simplicity we restrict the formal derivations to observables with discrete and nondegenerate spectra and to unitary repeatable measurements. The main results can then be demonstrated easily, without unnecessary mathematical complications. In the case of probability theorem I only, we provide a generalization of this result to continuous observables. For more general situations concerning also unsharp observables and arbitrary measurements, we refer to the literature [BLM 91, Mit 93].

3.2 The statistical interpretation

Here we distinguish three interpretations of quantum mechanics that connect the formal expressions of the theory with experimental results; they were mentioned already in chapter 1. The first two interpretations – the minimal interpretation and the realistic interpretation – are of different explanatory power and correspond to different kinds of measurement. The third one – the many-worlds interpretation – is of interest for the statistics but not invalidated by the objectification problem.

3.2(a) Minimal interpretation

The minimal interpretation I_M is the weakest possible interpretation of the quantum mechanical formalism. It is in the spirit of David Hume and has the positivism of Ernst Mach, and was advocated in particular by Niels Bohr. It avoids any statements about properties of individual objects and instead refers only to observational data, i.e., the results of measurement.

Let S be an object quantum system prepared in a pure state φ and $A = \sum a_i P[\varphi^{a_i}]$ a discrete observable with eigenvalues a_i and eigenstates

φ^{a_i}. We can then distinguish three postulates that constitute the *minimal interpretation.*

1. The calibration postulate (C_M). If φ is an eigenstate φ^{a_i} of A and the system possesses the value a_i of A, then a measurement of A leads to a pointer value Z_i of the measuring apparatus M, indicating that the system S had the eigenvalue $a_i = f(Z_i)$ and was in the state φ^{a_i} before the measurement, where $f(Z)$ is the pointer function. The postulate (C_M) represents the probability-free part I_M^0 of the minimal interpretation. In terms of quantum mechanical measurements, this means that the product state $\varphi^{a_i} \otimes \Phi$ of the preparation is transformed by a unitary operator U into a product state $\varphi_i' \otimes \Phi_i$, where the pointer state Φ_i indicates the pointer value Z_i. The object state φ_i' is a correlated but otherwise arbitrary state of S.

2. The pointer objectification postulate (PO). If φ is not an eigenstate of A, then it is not possible to predict with certainty the pointer value after the measurement of A. However, the postulate (PO) requires that after an A-measurement the pointer possesses some objective value Z_i even if it cannot predicted with certainty. According to the quantum theory of measurement, the state of an apparatus M after a unitary premeasurement is given by the reduced mixed state $W_M' = \sum |c_i|^2 P[\Phi]$, where the $|c_i|^2 = p(\varphi, a_i)$ are probabilities in the formal sense. The postulate (PO) then requires that the state W_M' admits an ignorance interpretation, which means that W_M' may be interpreted as a Gemenge $\Gamma(W_M')$, i.e., as a mixture $\{p(\varphi, a_i), \Phi_i\}$ of states with weights $p(\varphi, a_i)$. In this case, it is objectively decided which state Φ_i is actually realized even if this is subjectively unknown to the observer.

3. The probability reproducibility condition (PR). In addition to pointer objectification, the (PR) postulate requires that the formal probability $p(\varphi, a_i) = \text{tr}\{P[\varphi]P[\varphi^{a_i}]\}$, which is induced by φ and A, is reproduced in the statistics of the pointer values $Z_i^{(n)}$ after A-measurements on N identically prepared systems $S^n(\varphi)$ with $n = 1, 2, \ldots, N$. If the calibration postulate (C_M) is fulfilled, then for the operator U of a unitary premeasurement, $U(\varphi^{a_i} \otimes \Phi) = \varphi_i' \otimes \Phi_i$. By linearity, it then follows that for an arbitrary preparation φ the correlated compound state after the premeasurement reads $\Psi'(S + M) = U(\varphi \otimes \Phi) = \sum c_i \varphi_i' \otimes \Phi_i$ with $c_i = (\varphi^{a_i}, \varphi)$. Correspondingly, the state of the apparatus is given by $W_M' = \sum |c_i|^2 P[\Phi]$. The postulate ($PR$) then requires that the formal probabilities $p(\varphi, a_i) = |c_i|^2$ which appear in the spectral decomposition of W_M' are the probabilities of finding the pointer values Z_i after the A-measurement.

3.2(b) Realistic interpretation

In the case of repeatable measurements of the observable A, one can apply the stronger realistic interpretation I_R, which refers not only to the pointer values but also to the object system. Again one can distinguish three postulates.

1. The realistic calibration postulate (C_R). If φ is an eigenstate φ^{a_i} of A and the system possesses the value a_i of A, then an A-measurement leads with certainty to a pointer value Z_i that shows that S possesses the value a_i after the measurement. The postulate C_R represents the probability-free part I_R^0 of the realistic interpretation. In terms of measurements, this means that the product state $\varphi^{a_i} \otimes \Phi$ of the preparation is transformed by a unitary operator U into the product state $\varphi_i^a \otimes \Phi_i$ with pointer state Φ_i and the unchanged state φ^{a_i} of S. This means that the operator U provides a repeatable measurement.

2. The system objectification postulate (SO). If φ is not an eigenstate of A, then it is not possible to predict the A-value of S with certainty. However, the postulate (SO) requires that the system possess some objective value a_i of A after the measurement even if this cannot be predicted.

After a repeatable measurement, the state of the object system S is given by the reduced mixed state $W_S' = \sum |c_i|^2 P[\varphi^{a_i}]$ with the formal probabilities $p(\varphi, a_i) = |c_i|^2$. The postulate ($SO$) then requires that the state W_S' admits an ignorance interpretation, which means that W_S' may be interpreted as a mixture $\{p(\varphi, a_i), \varphi^{a_i}\}$ of states φ^{a_i} with weights $p(\varphi, a_i)$. It is then objectively decided which state φ^{a_i} pertains to S, but this is in general subjectively unknown to the observer.

3. The probability reproducibility condition (SR). In addition to the system objectification (SO), the postulate (SR) requires that the initial probability $p(\varphi, a_i)$ be not only reproduced in the statistics of the pointer values but also in the statistics of the A-values $a_i^{(n)}$ of the N object systems S_n after the measurement.

If the calibration postulate C_R is fulfilled for some operator U then $U(\varphi^{a_i} \otimes \Phi) = \varphi^{a_i} \otimes \Phi_i$. By linearity, it follows that for an arbitrary preparation φ the compound state after the premeasurement reads $\Psi'(S + M) = U(\varphi \otimes \Phi) = \sum c_i \varphi^{a_i} \otimes \Phi_i$. Hence the state of the object system is given by the mixed state $W_S' = \sum |c_i|^2 P[\varphi^{a_i}]$. The postulate ($SR$) then requires that the formal probabilities $p(\varphi, a_i) = |c_i|^2$ are the probabilities for finding the system values a_i after the A-measurement.

The two interpretations I_M and I_R of the quantum mechanical formalism obviously differ with respect to their explanatory power. Whereas in the minimal interpretation the theoretical terms refer only to pointer values Z_i, the realistic interpretation relates the theoretical terms also to A-values a_i of the object system. However, the realistic interpretation can only be applied to the formalism in the case of repeatable measurements, in contrast to the minimal interpretation the applicability of which is not restricted to a certain class of measurements.

3.2(c) The many-worlds interpretation

The third interpretation that is considered here is the many-worlds interpretation I_{MW}, which has already been discussed in detail in subsection 1.2(c). Here we confine our considerations to the interpretation of the probability structure. It should be emphasized that it was in the relative-state interpretation of Everett [Eve 57] and Wheeler [Whe 57] that the idea was first conceived that the probability structure of quantum mechanics evolves from the theory itself.

In its most general version the many-worlds interpretation is concerned with strong state-correlation measurements, and it is even weaker than the minimal interpretation, since pointer objectification is not required here. Hence we distinguish two versions of this interpretation, a 'minimal' version, dealing with unitary strong state-correlation premeasurements of sharp and discrete observables, and a 'realistic' interpretation, which is concerned only with repeatable measurements of sharp and discrete observables. For the sake of simplicity, we will discuss here only the realistic version of the many-world interpretation. The generalization to strong state-correlation measurements is then straightforward.

In order to characterize this realistic version of the many-worlds interpretation, we distinguish two postulates.

1. The many-worlds calibration postulate $(C_R)_{MW}$. If the preparation φ of S is the eigenstate φ^{a_i} of A and the system possesses the value a_i of A, then an A-measurement leads with certainty to a pointer value Z_i, showing that S possesses the value a_i after the measurement.

This postulate does not differ from the corresponding calibration postulate (C_R) of the realistic interpretation. The product state $\varphi^{a_i} \otimes \Phi$ of the preparation is transformed into the product state $\varphi^{a_i} \otimes \Phi_i$ after the measurement, and hence there is no need for further interpretation. However, if the preparation φ is not an eigenstate of A then neither the pointer objectification (PO) nor the system objectification (SO) is required here. This means that the post-measurement situation is completely described by the compound state $\Psi'(S + M) = U(\varphi \otimes \Phi) = \sum c_i \varphi^{a_i} \otimes \Phi_i$, with $c_i = (\varphi^{a_i}, \varphi)$. The meaning of these coefficients is then given by the many-worlds version of the probability reproducibility condition.

In the interpretations I_M and I_R the postulates (PR) and (SR), respectively, require that the coefficents $p(\varphi, a_i) = |c_i|^2$ be the probabilities for outcomes a_i in the reduced mixed states W'_S and W'_M of S and M, which are assumed to admit an ignorance interpretation, i.e., the objectification postulates (PO) and (SO) are presupposed. Here we dispense with the objectification require-ment, and hence the reduced mixed states have no independent empirical meaning.

This means that the 'measurement' is nothing else but the correlation between the state φ^{a_i} of the system and the 'relative state' Φ_i of the observer-apparatus, which is aware of the system state φ^{a_i}. The state $\Psi'(S + M)$ provides the state correlation between S and M, which can be considered already as a description of the complete measuring process. In this case too, the coefficients $p(\varphi, a_i)$ are probabilities in the formal sense. However, they cannot be interpreted here as relative frequencies of outcomes a_i in the Gemenge $\Gamma\{W'_S\} = \{p(\varphi, a_i), \varphi^{a_i}\}$, since this Gemenge plays no role in the description of the measuring process in the interpretation I_{MW}.

An interpretation of the coefficients $p(\varphi, a_i)$ can be given here in the following way. Let $S^{(N)}$ be a compound system of N identical, equally prepared systems with state φ. If for each system the observable A is measured the observer-apparatus will register a sequence $l = \{l_1, l_2, \ldots, l_N\}$ of N index values l_i that indicate the values a_{l_i} and store this in its memory. The formal probabilities $p(\varphi, a_i)$ will then be reproduced in the statistics of these memory sequences l. This means that for a given observer-apparatus the relative frequency $f^N(i, l)$ of the value a_i will approach, for large N, the probability $p(\varphi, a_i)$ in (almost) every memory sequence l. On the basis of these arguments one can now formulate the second postulate.

2. The many-worlds probability reproducibility condition $(SR)_{MW}$. For any def-inite observer represented by an apparatus M, the real positive number $p(\varphi, a_i)$ is the probability for the value a_i of A, which manifests itself as the relative frequency $f^N(i, l)$ of the value a_i in (almost) every memory sequence l of the observer-apparatus, in the limit $N \to \infty$.

The two postulates $(C_R)_{MW}$ and $(SR)_{MW}$ characterize the realistic version of the many-worlds interpretation.

3.3 Probability theorem I (minimal interpretation I_M)

At first glance, the postulates (C_M), (PO) and (PR) of the minimal interpretation I_M and the corrresponding postulates (C_R), (SO) and (SR) of the realistic interpretation I_R seem to be independent requirements that characterize the respective interpretations I_M and I_R. However, it turns out that the conditions (PR) and (SR), i.e., the probabilistic parts of the interpretations I_M and I_R, follow from *probability-free interpretations*, I_M^0 and I_R^0, respectively. This important and somewhat surprising result is the content of probability theorems I and II, to be discussed in the present section and in the next section.

3.3(a) Discrete observables

Within the interpretation I_M, the calibration postulate C_M corresponds to a probability-free minimal interpretation I_M^0, which merely states that for a system that possesses the value a_i of A, a measurement of A leads with certainty to the pointer value $Z_i = f^{-1}(a_i)$, which is correlated with the value a_i. The quantum theory of measurement provides explicit examples of unitary operators U_A that fulfil the requirement of the postulate C_M.

The calibration postulate C_M of the interpretation I_M implies that the post-measurement state of the apparatus M reads

$$W_M' = \sum |(\varphi^{a_i}, \varphi)|^2 P[\Phi_i].$$

However, the meaning of the coefficients $c_i = (\varphi^{a_i}, \varphi)$ is still open. It is easy to show that the real positive numbers $p(\varphi, a_i) = |(\varphi^{a_i}, \varphi)|^2$ are probabilities in the formal sense. The interpretation of the probability distribution $p(\varphi, a_i)$ can then be obtained from the probability reproducibility condition (PR).

According to this condition (PR), the probability distribution $p(\varphi, a_i)$ induced by φ and A is reproduced in the statistics of the pointer values Z_i. In other words, the number $p(\varphi, a_i)$ is the probability of finding after the measuring process the pointer value Z_i indicating that the value $a_i = f(Z_i)$ was measured. This means that if one were to perform a series of N measurements of the observable A on equally prepared systems S_n ($n = 1, 2, \ldots, N$), then in the limit $N \to \infty$ the relative frequency $f^N(\varphi, Z_i)$ of the pointer values Z_i would approach in some sense the probability $p(\varphi, a_i)$.

The probability reproducibility condition (PR), i.e., the probabilistic part of the minimal interpretation I_M, follows from the calibration postulate C_M, i.e., from the probability-free interpretation I_M^0, by means of quantum mechanics, provided the latter is assumed to be universally valid. In particular, it must be presupposed that quantum mechanics applies equally to single-object systems as well as to compound systems $S^{(N)}$ consisting of N equally prepared systems S_i.

In order to derive this interesting result, we consider N independent systems S_i, with equal preparations $\varphi^{(i)}$, as a compound system $S^{(N)}$ in the tensor product state $(\varphi)^N$. For each system S_i a premeasurement of A transforms the initial state Φ of M into the mixed state $W'_M = \sum p(\varphi, a_i) P[\Phi_i]$, with eigenstates Φ_i of the pointer observable Z corresponding to pointer values Z_i. If the observable A is measured for each system S_i, then the full measurement outcome is given by a sequence of pointer values $\{Z_{l_i}\}$ and pointer states $\{\Phi_{l_i}\}$, respectively, with an index set $l = \{l_1, l_2, \ldots, l_N\}$. In the N-fold tensor-product Hilbert space $H_M^{(N)}$ of the apparatus, the states $\Phi_l^N = \Phi_{l_1} \otimes \Phi_{l_2} \otimes \cdots \otimes \Phi_{l_N}$ form a complete orthonormal basis. For any product state Φ_l^N, the relative frequency $f^N(l, k)$ of the pointer value Z_k is objectively determined, and one can define an observable 'relative frequency of the pointer value Z_k' by $F_k^N = \sum_l f^N(k, l) P[\Phi_l^N]$, where the sum runs over all sequences l.

After a premeasurement of A, the apparatus is in the mixed state W'_M. The sequence of N measurements is then described by the N-fold product $(W'_M)^N$ of these mixed states. In general, this product state is not an eigenstate of the relative frequency operator F_k^N, which means that the relative frequency of the pointer value Z_k is not an objective property in the mixed state $(W'_M)^N$. However, in the limit of large N the post-measurement product state $(W'_M)^N$ becomes an eigenstate of the relative-frequency observable F_k^N, and the relative frequency of the pointer value Z_k approaches the probability $p(\varphi, a_k)$. This is the content of probability theorem I, which can formally be expressed as follows.

Probability theorem I

$$\lim_{N \to \infty} \mathrm{tr}\left\{ (W'_M)^N (F_k^N - p(\varphi, a_k))^2 \right\} = 0.$$

It holds in the minimal interpretation I_M (see Appendix 1).

3.3(b) Random sequences

For a given index sequence $l = \{l_1, l_2, \ldots, l_N\}$ with pointer values $\{Z_{l_i}\}$ and system values $\{a_{l_i}\}$, $a_{l_i} = f(Z_{l_i})$, the relative frequency $f^N(k, l)$ will approach the probability $p(\varphi, a_k)$ in the described sense only if $l = \{l_i\}$ is a random sequence. In order to measure the degree to which a sequence l deviates from a random sequence with weights $p(\varphi, a_i)$, we define the function

$$\delta(l) := \sum_k (f^N(k, l) - p(\varphi, a_k))^2,$$

which is the first in a hierarchy of functions of this kind. A sequence will then be called *first random* if $\delta(l) < \varepsilon$ holds for an arbitrary small positive number ε.

A unitary premeasurement of one system $S_i(\varphi)$ with an initial pointer state $\Phi^{(i)}$ leads to the compound state

$$\Psi'(S + M) = U(\varphi^{(i)} \otimes \Phi^{(i)}) = \sum_k c_k \varphi_k^{(i)} \otimes \Phi_k^{(i)},$$

where we have assumed (for the sake of simplicity) a repeatable premeasurement. If a measuring process is performed with N equally prepared systems, we have to consider the final state

$$\Psi^{(N)'} = \bigotimes_{i=1}^N \left(\sum_k c_k \varphi_k^{(i)} \otimes \Phi_k^{(i)} \right) = \sum_l c_{\{l\}} (\varphi)_l^N \otimes \Phi_l^N$$

where $c_{\{l\}} = c_{l_1} c_{l_2} \cdots c_{l_N}$, $(\varphi)_l^N = \varphi_{l_1}^{(1)} \otimes \varphi_{l_2}^{(2)} \otimes \cdots \otimes \varphi_{l_N}^{(N)}$, and $\Phi_l^N = \prod_i \Phi_{l_i}^{(i)}$. If we remove from this superposition all sequences l that are non-first-random we obtain

$$\Psi_\varepsilon^{(N)'} = \sum_{l, \delta < \varepsilon} c_{\{l\}} (\varphi)_l^N \otimes \Phi_l^N$$

and the difference is given by

$$\chi_\varepsilon^N := \Psi^{(N)'} - \Psi_\varepsilon^{(N)'} = \sum_{l, \delta \geq \varepsilon} c_{\{l\}} (\varphi)_l^N \otimes \Phi_l^N.$$

Using some results for the relative frequency function $f^N(k, l)$ one can show that (Appendix 2)

$$(\chi_\varepsilon^N, \chi_\varepsilon^N) \leq \frac{1}{N\varepsilon} \sum p(\varphi, a_k)[1 - p(\varphi, a_k)] \leq \frac{1}{N\varepsilon}$$

which means that in the limit $N \to \infty$ the contribution of the non-first-random sequences becomes arbitrarily small and $\Psi_\varepsilon^{(N)'}$ approaches $\Psi^{(N)'}$ [DeW 71].

3.3(c) Continuous observables

Probability theorem I can be applied also in the case where A is a continuous observable,

$$A = \int_{\mathbb{R}} a \, dP^A(a), \quad a \in \mathbb{R}.$$

In order to demonstrate this result, we first discretize the observable A, making use of a fixed partition (X_i) of the value set $X^A = \mathbb{R}$ of the system into mutually disjoint Borel subsets $X_i \in \mathscr{B}(X^A)$. There are uncountably many partitions of this kind labelled by some index value $\lambda \in \mathbb{R}$. Any partition $(X_i)_\lambda$ induces a discrete A-observable $A^\lambda : i \rightarrow P^A(X_i^\lambda)$ and a reading scale R_λ defining a discretized version $Z^\lambda : i \rightarrow Z_i^\lambda = f^{-1}(X_i^\lambda)$ of the pointer observable Z. The total measurement probability $p^A(\varphi, a), a \in \mathscr{B}(X^A)$, can then be recovered for every state φ from the probabilities $p^{A^\lambda}(\varphi, X_i^\lambda)$ for the discretized observables A^λ, by varying over all possible reading scales R_λ.

For any fixed partition (X_i^λ) of X^A, probability theorem I can be applied. This means that the probability $p^{A^\lambda}(\varphi, X_i^\lambda)$ that is induced by A^λ and φ is reproduced in the statistics of the coarse-grained version Z^λ of the pointer observable. The relative frequency $f^N(Z_i^\lambda, l)$ of the pointer value Z_i^λ in a sequence labelled by l of N measurements will then approach the probability $p^{A^\lambda}(\varphi, X_i^\lambda)$ for increasing N. In this way, for each reading scale R_λ it can be demonstrated that the probability distribution $p^{A^\lambda}(\varphi, X_i^\lambda)$ is reproduced in the statistics of the pointer outcomes Z_i^λ given by the relative frequencies $f^N(Z_i^\lambda, l)$ of the outcome sequences l.

The full measurement statistics $\{f^N(Z, l)\}_{N \rightarrow \infty}$, $Z \in f^{-1}(X^A)$, can then be obtained by varying the coarse-grained statistics $\{f^N(Z_i^\lambda, l)\}_{N \rightarrow \infty}$ over the whole family $\{R_\lambda\}$ of reading scales. The total measurement probability $p^A(\varphi, a)$, $a \in \mathscr{B}(X^A)$, which is obtained from the probabilities $p^{A^\lambda}(\varphi, X_i^\lambda)$ by varying over all reading scales $\{R_\lambda\}$, is then reproduced in the full measurement statistics $\{f^N(Z, l)\}_{N \rightarrow \infty}$. In this way, the demonstration of the probability reproduction in the statistics of the pointer values, which was formulated for discrete observables in probability theorem I, can be extended also to continuous observables.

3.4 Probability theorem II (realistic interpretation I_R)

3.4(a) Discrete observables

In the case of repeatable measurements, one can apply the realistic interpretation I_R. The calibration postulate C_R corresponds again to the probability-free part I_R^0 of the realistic interpretation. It requires that, for a system S which possesses a value a_i of A, a measurement of A leads not only with certainty to a pointer value Z_i which indicates the result a_i but also to an object value a_i, i.e., the measurement preserves the value a_i of the object system. Clearly, this requirement can only be fulfilled by repeatable measurements.

The calibration postulate C_R implies that the post-measurement state of the object system reads $W_S' = \sum p(\varphi, a_i) P[\varphi^{a_i}]$, where the $p(\varphi, a_i)$ are again the initial probabilities induced by the preparation φ and the observable A. The interpretation of the formal probabilities is given here by the probability reproducibility condition (SR): $p(\varphi, a_i)$ is the probability that after the measurement of A the system S possesses the value a_i of A. This means that if one were to perform A-measurements on N equally prepared systems $S_n(\varphi)$ the relative frequency $f^N(\varphi, a_i)$ of object systems $S_n(\varphi)$ with A-values a_i would approach in some sense the probability $p(\varphi, a_i)$ for $N \to \infty$.

In order to derive this result one must again assume that quantum mechanics can be applied not only to a single object system S_i but also to a compound system $S^{(N)}$ that consists of N equally prepared systems S_i. The totality $\{S_i\}^N$ of N systems can then be considered as a compound system $S^{(N)}$ in the tensor product state $(\varphi)^N$. A premeasurement of A transforms the initial state $\varphi^{(i)}$ of each system S_i into the mixed state $W_S' = \sum p(\varphi, a_i) P[\varphi^{a_i}]$, with eigenstates φ^{a_i} of A corresponding to the values $a_i = f(Z_i)$. If A is measured for each system S_i, the measurement result is given by a sequence $\{a_{l_1}, a_{l_2}, \ldots, a_{l_N}\}$ of object values a_{l_i} and states $\varphi^{a_{l_i}}$ respectively, with index set $l = \{l_1, l_2 \ldots, l_N\}$.

In the N-fold tensor-product Hilbert space $H_S^{(N)}$ of the compound system $S^{(N)}$ the states $(\varphi)_l^N = \varphi_{l_1}^{(1)} \otimes \varphi_{l_2}^{(2)} \otimes \cdots \otimes \varphi_{l_N}^{(N)}$ with an index set l form a complete orthonormal basis. In a state $(\varphi)_l^N$, each subsystem S_i has a well-defined A-value a_i. The relative frequency $f^N(k, l)$ of a value a_k in the state $(\varphi)_l^N$ is then objectively determined and we can define an operator 'the relative frequency of systems with the value a_k' by $f_k^N = \sum_l f^N(k, l) P[(\varphi)_l^N]$, where the sum runs over all sequences l.

After a premeasurement, a system S_i is in a mixed state W_{S_i}'. If N premeasurements of A are performed, the state of the compound system $S^{(N)}$ is given by the N-fold tensor product state $(W_S')^N = W_{S_1}' \otimes W_{S_2}' \otimes \cdots \otimes W_{S_N}'$ of these mixed states W_{S_i}'. In general, this product state is not an eigenstate of the relative frequency operator f_k^N, and thus the relative frequency of the values a_k is not an objective property of the system $S^{(N)}$ in the state $(W_S')^N$. However, in the limit of large values of N the post-measurement product state of the compound system becomes an eigenstate of f_k^N and the value of the relative frequency of a_k approaches the probability $p(\varphi, a_k)$. This is the content of

Probability theorem II

$$\lim_{N \to \infty} \text{tr}\left\{ (W_S')^N (f_k^N - p(\varphi, a_k))^2 \right\} = 0.$$

Its validity is restricted to the realistic interpretation I_R (Appendix 3).

3.4(b) Additional remarks

Probability theorem II is formulated for discrete, nondegenerate observables. There is no problem in extending the theorem to degenerate observables if Lüders measurements are used. As pointed out above (2.3(c)), Lüders measurements are repeatable and thus admit the realistic interpretation I_R. The proof of probability theorem II as presented in Appendix 3 can easily be generalized for the case of a degenerate observable A.

However, it is not possible to extend this theorem to the case of a continuous observable. The reason is that a continous observable does not admit repeatable measurements. More precisely, according to the theorem of Ozawa [Oza 84] mentioned above (2.4(c)), we know that if an observable A admits a repeatable measurement, then A is discrete. Consequently, if an observable A is continuous, then there is no repeatable measurement, and the realistic calibration postulate (C_R) cannot be fulfilled. Hence the initial probability $p(\varphi, a)$ is no longer expressed in the post-measurement object state W_S' and reproduced in the statistics of the post-measurement system values.

3.5 Probability theorems III and IV

3.5(a) Remarks on the work of Everett, Graham, Finkelstein, and Hartle

There are some further approaches to justifying the statistical predictions of quantum mechanics. However, they differ in an essential point from the arguments used in probability theorems I and II. Instead of referring the statistical predictions to the situation after the measuring process, and hence to the post-measurement states of the object or the apparatus, these attempts understand – for varying reasons – the probabilistic propositions as statements about the situation prior to the measurement.

The first attempt to justify the statistical interpretation of quantum mechanics on the basis of quantum-theoretical statements about individual systems was made by Everett [Eve 57] in 1957 within the framework of the many-worlds interpretation mentioned above (cf. 3.2(c)). However, Everett did not formulate this new result as a theorem, nor did he provide arguments that could serve as a convincing proof. Within the context of the many-worlds interpretation, a theorem of this kind together with its proof was first communicated by Graham [Gra 73] and further improved by DeWitt [DeW 71]. Clearly, these authors related the statistical predictions of quantum mechanics to the initial preparation $\varphi \otimes \Phi$ of the object system and the apparatus, since according to the many-worlds interpretation this state is never changed by objectification and reading but persists – except for unitary time development.

In 1968, Hartle [Hart 68] published a paper with the title 'Quantum mechanics of individual systems'. In this article, he presented a formulation of quantum mechanics that begins with assertions for individual systems and derives the statistical predictions of the theory. In particular, Hartle succeeded in showing that quantum probabilities are relative frequencies in an ensemble of N equally prepared systems $S_i(\varphi)$ provided that the number N is sufficiently large. In contrast to probability theorem II, but as in the approach of Everett and Graham mentioned above, Hartle did not refer the statistical predictions to the reduced state of the object system after the measurement but proved the equivalence of probability propositions (about S_i) and 'yes–no' propositions (about $S^{(N)}$) for the pure-state preparation prior to the measurement. From a technical point of view, Hartle's result is similar to that of Graham, which was obtained independently. It is, however, more rigorous and deals carefully with the problem of defining the Hilbert space in the limit of infinitely many systems. In addition, the treatment is completely independent of the strange ontological implications of the many-worlds interpretation. The theorem of Hartle and its proof will be presented in the next subsection (3.5(b)).

A few years earlier than Hartle, and again independently of Everett, Finkelstein [Fink 62] investigated the statistical problem from a somewhat different angle. Being interested primarily in quantum logic, Finkelstein tried to show that quantum mechanics may be considered as a theory of yes–no propositions. For this reason, he formulated the probabilistic statements of quantum mechanics as yes–no propositions that refer to an ensemble of equally prepared systems and indicated a proof of this equivalence. Finkelstein did not refer to the minimal interpretation and the measuring process. Instead, he formulated his results with respect to the preparation states of single systems and of the ensemble.

In this paper, 'The logic of quantum physics' [Fink 62], Finkelstein considered the following problem: for a single quantum system S with preparation φ one can formulate, as in classical physics, yes–no propositions that refer to the objective properties of the system $S(\varphi)$. In addition, in quantum mechanics there are also probability statements $p^A(\varphi, a_i)$ that are concerned with the nonobjective properties of the system $S(\varphi)$. However, according to Finkelstein these probability statements may be considered as yes–no propositions the referent of which is an ensemble $S^{(N)}$ of N equally prepared systems $S(\varphi)$, provided that the number N is sufficiently large.

3.5(b) Formulation of the theorems and proofs

(i) Finkelstein's theorem

In order to justify the conjecture that probability statements may be considered as yes–no propositions that refer to an ensemble of equally prepared systems, one can argue in the following way. Let $\{S_i(\varphi)\}$ be a set of identical quantum systems S_i with equal preparations $\varphi^{(i)} \in \mathcal{H}_i$, where $\mathcal{H}_i$ is the Hilbert space of S_i. The compound system $S^{(N)} = S_1 + S_2 + \cdots + S_N$ can then be described by the tensor product state $(\varphi)^N = \varphi^{(1)} \otimes \varphi^{(2)} \otimes \cdots \otimes \varphi^{(N)}$ with $\varphi^{(N)} \in \mathcal{H}_{(N)}$. For an observable $A = \sum a_i P[\varphi^{a_i}]$ of a single system S one can define a new N-body observable in $\mathcal{H}^{(N)}$, the *mean value of A* by

$$\bar{A} = \frac{1}{N}(A^{(1)} + A^{(2)} + \cdots + A^{(N)})$$

where $A^{(k)} = A_k \overset{N}{\underset{i \neq k}{\otimes}} \mathbb{1}_k$, and A_k is the A-observable of the system S_k with Hilbert space $\mathcal{H}_k$. If one defines

$$\Delta^2(N, \xi) := \|\bar{A}^N(\varphi)^N - \xi(\varphi)^N\|^2, \quad \xi \in \mathbb{R}$$

one obtains after tedious but straightforward calculations (Appendix 4)

Lemma 3.1

$$\Delta^2(N, \xi) = \frac{N-1}{N}\{(\varphi, A\varphi) - \xi\}^2 + \frac{1}{N}(\varphi, (A - \xi)^2\varphi),$$

which means that Δ^2 is a sum of two positive terms $\frac{N-1}{N}\Delta_1^2$ and $\frac{1}{N}\Delta_2^2$. Using this result one can argue in the following way.

1. If $\lim\limits_{N \to \infty} \Delta^2(N, \xi) = 0$ then Δ_1^2 must vanish and thus we have $\xi = (\varphi, A\varphi)$.
2. If $\xi = (\varphi, A\varphi)$ then $\Delta_1^2 = 0$ and $\Delta^2 = \frac{1}{N}\Delta_2^2$ and thus $\lim\limits_{N \to \infty} \Delta^2(N, \xi) = 0$.

On the basis of these arguments we can now formulate

Probability theorem III (Finkelstein 1962). *Let*

$$\Delta^2(N, \xi) := \|\bar{A}^N(\varphi)^N - \xi(\varphi)^N\|^2;$$

then $\lim_{N \to \infty} \Delta^2(N, \xi) = 0$ *if and only if* $\xi = (\varphi, A\varphi)$.

The interpretation of this theorem leads to the following result. The observable $\bar{A}^N$: 'mean value of A' of the compound system $S^{(N)}$ possesses objectively the value $(\varphi, A\varphi)$ in the limit $N \to \infty$. Even if A is not an objective observable of an individual system $S(\varphi)$, the observable $\bar{A}^N$ pertains to the compound system $S^{(N)}$ in state $(\varphi)^N$ as an objective property with the value $(\varphi, A\varphi)$. This means that the probabilistic statements about nonobjective properties of $S(\varphi)$ are yes–no propositions that refer to the compound system $S^{(N)}$. Consequently, quantum mechanics can be understood as a theory which makes use only of yes–no propositions that refer partly to an individual systems and partly to an ensemble.

(ii) Hartle's theorem

In order to formulate Hartle's theorem, we adopt the terminology of probability theorem II. If the individual systems S_i have the preparations $\varphi^{(i)} = \varphi$, then the state of the compound system $S^{(N)}$ is given by the tensor product state $(\varphi)^N$. Using again the operator 'relative frequency of the systems with value a_k', which is given by $f_k^N = \sum_l f^N(k, l) P[(\varphi)_l^N]$ with states

$$(\varphi)_l^N = \varphi_{l_1}^{(1)} \otimes \varphi_{l_2}^{(2)} \otimes \cdots \otimes \varphi_{l_N}^{(N)}$$

and relative frequencies $f^N(k, l)$ of the index k in the index sequence l of length N, Hartle's result reads:

Probability theorem IV (Hartle 1968)

$$\lim_{N \to \infty} \mathrm{tr}\left\{ P[(\varphi)^N](f_k^N - p(\varphi, a_k))^2 \right\} = 0.$$

This theorem states that for a compound system $S^{(N)}$ with state $(\varphi)^N$ the relative frequency of the A-value a_k is an objective property of $S^{(N)}$ and is given by the value $p(\varphi, a_k)$, provided that the number N of systems $S_i(\varphi)$ is sufficiently large.

For the proof of this result, we note that Hartle's theorem is a special case of probability theorem II, the proof of which applies also here. Indeed, the reduced mixed state W_S' of the object system that appears in probability theorem II is given by

$$W'_S(\varphi, U) = \text{tr}_M\{P[U(\varphi \otimes \Phi)]\}.$$

For the special case of vanishing object–apparatus interaction, i.e., for $U = 1$, we obtain

$$W'_S(\varphi, U) = P[\varphi]$$

and thus $(W'_S)^N = P[(\varphi)^N]$. Inserting this special reduced state into the formula of probability theorem II we obtain Hartle's theorem.

Moreover, it is interesting to note that Hartle's theorem follows also from Finkelstein's theorem and can thus be proved in this way. The implication between the two theorems can easily be seen (Appendix 5).

The meaning of Hartle's theorem is not quite clear. None of the individual systems $S_i(\varphi)$ with state φ possesses an A-value a_i in an objective sense. According to the assumptions made, the observable A is nonobjective with respect to the state φ. In spite of this obvious situation, Hartle's theorem claims that in the tensor product state $(\varphi)^N$ the relative frequency of a value a_k is an objective property of the compound system $S^{(N)}$, provided that the number N is sufficiently large. In the case of probability theorem II of the realistic interpretation, the situation was at first similar. The post-measurement states W'_S of the individual systems are not eigenstates of A, and without further assumptions none of the systems possesses a value a_i. However, if the states W'_S admit an ignorance interpretation, i.e., if the system objectification (SO) is postulated, then each system possesses objectively an A-value, and the relative frequency of the values a_i in the compound system $S^{(N)}$ becomes a meaningful concept.

There are at least two fields of application for Hartle's theorem. Hartle's original intention was to elaborate a quantum theory of individual systems. If there is only one quantum system in the world and no external measuring apparatus exists, what could be the meaning of the state φ of this system? Probability theorem IV gives an answer to this question and shows that the formal expressions $p(\varphi, a_i)$ have a meaning as properties of an ensemble $S^{(N)}$ of equally prepared systems. For an individual system this statement is rather irrelevant since there is no ensemble $S^{(N)}$. However, the theorem shows that for the interpretation of the wave function one does not need an apparatus that provides a 'collapse' of the state.

These results become relevant if one investigates actual situations with only one individual system, e.g. quantum cosmology. For the early stages of the universe, it is no longer sufficient to consider the dynamics of cosmological models within the framework of general relativity; quantum theory must also be taken into account. For this reason, it is sometimes meaningful to study the 'wave function of the universe'. Clearly, there is no additional external system that could play the role of an observer-apparatus. Moreover, it is completely impossible to influence the system dynamically so that the state of the universe is changed by a collapse of the wave function. It is obvious that Hartle's result can be applied to problems of this kind and there are indeed several investigations that are concerned with this topic. We mention here, e.g., the papers by Hartle and Hawking [HaHa 83] and by Gell-Mann and Hartle [GeHa 90].

Everett was the first who considered the possibility of applying quantum theory to the universe as a whole. In particular, he showed how·an observer-apparatus could be incorporated and how its activities – measuring and recording – could be described by quantum mechanics. His result, the extension of the realistic interpretation to individual systems, is the 'relative-state' formulation of quantum mechanics, which became known later as the many-worlds interpretation (cf. 1.2(c) and 3.2(c)). Within the framework of this interpretation, a result similar to that of Hartle was obtained by Graham [Gra 73]. Hence it is obvious that Hartle's theorem, which is a mathematical improvement of Graham's investigations, applies also to the many-worlds interpretation.[†]

In this sense, probability theorem IV provides a justification of the many-worlds version of the probability reproducibility condition, $(SR)_{MW}$ of subsection 3.2(c). In comparison with the derivation of the probability reproducibility conditions (PR) and (SR) by means of probability theorems I and II respectively, the present justification is even more rigorous. For the interpretation of probability theorems I and II, the objectification postulates (PO) and (SO) must be presupposed. (This problem will be discussed in subsection 3.6(a).) Since the many-worlds interpretation is an interpretation without objectification, for the justification of condition $(SR)_{MW}$ nothing but the calibration postulate $(C_R)_{MW}$ must be assumed.

3.6 Interpretation

3.6(a) Probability theorems and objectification

For the *derivation* of probability theorem I, one must assume that the calibration postulate (C_M) holds generally for single systems and for compound systems. For *interpretation* of the theorem as a justification of the statistical

---

† It should be mentioned that Hartle's original paper was not concerned with the many-worlds interpretation but presented a mathematical result on individual systems.

(Born) interpretation of quantum mechanics, this premise is not sufficient. In addition, one must presuppose that after each measurement of A the pointer possesses objectively a value $Z_l^{(i)}$ such that the relative frequency of some value Z_k can be directly obtained from these results. Indeed, probability theorem I states that in the limit $N \to \infty$ the relative frequency of a value Z_k is an objective property of the ensemble of post-measurement pointer states. This means that the reduced mixed states $W_M^{(i)'}$ of the measuring apparatus after the measurement must admit an ignorance interpretation as a Gemenge $\Gamma(W_M^{(i)'}) = \{p(\varphi, a_k), Z_k^{(i)}\}$, i.e., as a mixture of states $\Phi_k^{(i)}$ or of pointer values $Z_k^{(i)}$ with statistical weights $p(\varphi, a_k)$. Obviously, this assumption is simply the pointer objectification postulate (PO), which must thus be presupposed for the interpretation of probability theorem I.

An equivalent situation arises for probability theorem II. Again, the formal validity is guaranteed if the realistic calibration postulate (C_R) holds both for individual systems and for compound systems. For the interpretation of probability theorem II as a justification of the statistical interpretation of quantum mechanics one has to presuppose that the states W_{S_i}' of the object system after the measuring process admit an ignorance interpretation such that a system S_i possesses objectively the A-value $a_k^{(i)}$ with statistical weight $p(\varphi, a_k)$. Otherwise, one could not determine the relative frequency of the value a_k by counting the measurement outcomes in the ensemble $S^{(N)}$. Probability theorem II states that the relative frequency of the value a_k is an objective property of the ensemble $S^{(N)}$ after the measurement, with the limiting value $p(\varphi, a_k)$. It is obvious that the verifiability of this statement presupposes the possibility of objectifying the A-values of the systems S_i in the states W_{S_i}', i.e., it presupposes the system objectification postulate (SO).

These arguments show that, on the one hand, for the interpretation of probability theorems I and II the pointer objectification or the system objectification, respectively, are necessary preconditions. On the other hand, it will be shown in the next chapter that neither the system objectification postulate (SO) nor the pointer objectification postulate (PO) can be justified within the quantum theory of unitary measurements. However, since the formal derivation of probability theorems I and II has just been performed within the framework of the theory of unitary premeasurements, we arrive at a somewhat confusing result: probability theorems I and II can only be proved under conditions (unitary premeasurements) that exclude their interpretation as a justification of the probability interpretation of quantum mechanics.

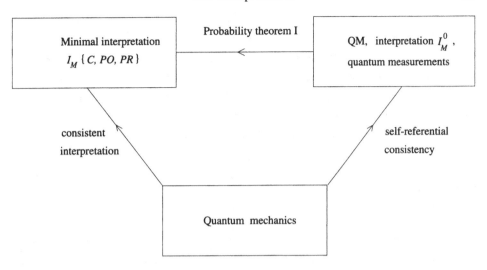

Fig. 3.1 Self-referential consistency between quantum object theory and its minimal interpretation I_M with respect to the (PR) postulate.

3.6(b) Self-referentiality

Probability theorems I and II provide some kind of self-referential consistency of quantum mechanics. In the most general case, the empirical meaning and the possibility of an experimental verification of the theory are given by the minimal interpretation I_M. In particular, this means that the theoretical probability expressions $p^A(\varphi, a_i)$ are reproduced in the statistics of the observable pointer values Z_i. Probability theorem I shows that this probability reproducibility condition is not an additional requirement that must be added to the theory. Instead, it turns out that this result can be obtained by means of the quantum theory of measurement and the probability-free minimal interpretation I_M^0. However, since the quantum theory of measurement and the calibration postulate follow from quantum mechanics, it follows that quantum mechanics implies the statistical part of exactly that interpretation I_M which must be presupposed in any interpretation of the theory.

In this way, one obtains some self-referential consistency between quantum object theory and its minimal interpretation I_M – at least with respect to the probability reproducibility condition (PR) – in the following sense. On the one hand, for the statistical interpretation of the theoretical terms $p^A(\varphi, a_i)$ one needs the condition (PR); but this – on the other hand – can be derived from the quantum mechanical formalism and the probability-free part I_M^0 of the minimal interpretation. This self-referential consistency is illustrated schematically in Fig. 3.1.

It should be noted that the relevance of this self-referential consistency is somewhat inhibited by the pointer objectification problem mentioned above. Since pointer objectification (PO) cannot be derived within the quantum theory of unitary measurements, for the interpretation of probability theorems I and II of the quantum mechanical formalism this postulate must therefore be added as an assumption that is inevitable but presently not provable.

In the case of repeatable measurements, one can apply the realistic interpretation I_R, which means, in particular, that the formal probabilities $p^A(\varphi, a_i)$ are also reproduced in the statistics of the A-values a_i of the object system. Probability theorem II then shows that the probability reproducibility condition (SR) is again not an additional postulate but a proposition that can be derived by means of the quantum theory of measurement and the probability-free realistic interpretation I_R^0. Hence we are confronted with a similar situation as for the minimal interpretation: quantum mechanics implies the statistical part (SR) of the same realistic interpretation that is used for the interpretation of quantum theory in the case of repeatable measurements.

This means that also here we obtain a self-referential consistency between the theory and its realistic interpretation. On the one hand, for the realistic statistical interpretation of the theoretical terms $p^A(\varphi, a_i)$ one applies the probability reproducibility condition (SR), but this is – on the other hand – derivable from the quantum theory of repeatable measurements and the probability-free part I_R^0 of the realistic interpretation. This self-referential consistency is illustrated schematically in Fig. 3.2.

The relevance of this self-referential consistency is again restricted by the fact that for repeatable measurements the system objectification postulate (SO) cannot be derived within the quantum theory of unitary measurements. Since the condition (SO) is necessary for the interpretation of theorem II and the quantum mechanical formalism, it must be added as an assumption that is presently not provable.

3.6(c) Individualistic interpretation of quantum mechanics

Probability theorems I and II and their extensions to continuous observables have important consequences for the problem of an individualistic interpretation of quantum mechanics. In order to demonstrate these implications let us briefly recall the usual probabilistic interpretation. For a system S with Hilbert space $\mathcal{H}_S$ and a maximal preparation $\varphi \in \mathcal{H}_S$, the discrete observables $A = A^+$ with $[A, P[\varphi]] = 0$ possess objective values $a_i \in X^A$, where $X^A = \{a_i\}$ is the value space of A. All the other observables

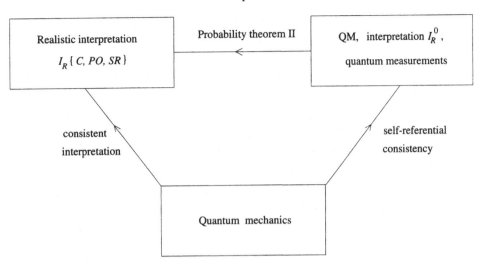

Fig. 3.2 Self-referential consistency between quantum object theory and its realistic interpretation I_R with respect to the (SR) postulate.

$B = \sum b_k P(b_k)$, with $\forall_k [P(b_k), P[\varphi]] \neq 0$, have no objective values $b_k \in X^B$ in the state φ. However, to the system $S = S(\varphi)$ one can attribute probabilities $p(\varphi, b_k) = \text{tr}\{P[b_k]P[\varphi]\}$ that provide some additional information about the values b_k after a measurement of the observable B. If the measurement is repeatable, the system S will possess some value b_k after the measurement, otherwise the pointer will have a value $Z_k = f^{-1}(b_k)$ that indicates the result b_k. The experimental verification of these probability propositions is a statistical one: if one were to perform a large number N of B-measurements on equally prepared systems $S(\varphi)$, then the relative frequency f_k^N of the outcome b_k would approach in some sense the value $p(\varphi, b_k)$ in the limit $N \to \infty$.

This means that a probability proposition $P\{p(\varphi, b_k)\}$ is a proposition the referent of which is a compound quantum system $S^{(N)}$ composed of N originally equally prepared systems $S_n(\varphi)$ after a measurement of B. The proposition $P\{p(\varphi, b_k)\}$ states that the relative frequency of obtaining the result b_k in the ensemble $S^{(N)}$ approaches the value $p(\varphi, b_k)$ for $N \to \infty$. This is the content of the probability reproducibility postulates (PR) and (SR), relating to the interpretations I_M and I_R respectively.

However, this explanation does not answer the question what the term $p(\varphi, a_i)$ could mean for an individual system S with preparation φ. This question is motivated by the fact that quantum probabilities express some genuine indeterminism and cannot be reduced to a distribution of hidden initial conditions. Clearly, a system $S(\varphi)$ possesses objectively values of all observables A with $[A, P[\varphi]] = 0$. On the other hand, probability statements concerning values of nonobjective properties can neither be verified nor falsified by experimental means since a single system can only be measured once.

In order to attach probabilities $p(\varphi, a_i)$ to a single physical system, two of-ten discussed proposals have been made by Heisenberg and by Popper. In the first approach, Heisenberg [Heis 59] referred to the philosophy of Aristotle. Being aware that quantum probabilities must not be considered as the quan-tification of the observer's subjective ignorance only, he tried to understand probabilities also as some kind of physical reality that pertains to the object system in question. In this sense, he argued that quantum probabilities are a quantitative formulation of the concept of *potentia* in Aristotle's philosophy. Accordingly, in quantum mechanics one should interpret the 'possibility' or 'tendency' that is quantified by $p(\varphi, a_i)$ as some intermediate layer of reality, halfway between the objective reality of quantum systems and the subjective reality of the observer's knowledge.

The second attempt to give some meaning to the probability of a single event was made by Popper [Pop 57]. He introduced the concept of *propensity* in order to provide a single-case interpretation of probability. However, the propensity concept was not primarily conceived as a means for interpreting quantum mechanics. Instead, it was understood by Popper as a new concept that can be used for all kinds of probabilistic problems: thus there is a subjectivistic interpretation, which is without interest for our problems, and an objectivistic interpretation, which might be of interest for physics. Within the objectivistic approach, one has to distinguish the deterministic case, e.g. dice-tossing experiments, and the indeterministic case, which presumably incorporates quantum mechanical situations. Whereas for deterministic situ-ations the propensity concept is often considered to be void [Skl 70, Gie 73, Schn 94], some authors claim that propensities might be particulary useful in quantum mechanical situations, which are considered to be indeterministic [Pop 82].

In spite of these intuitively rather convincing arguments, in an actual experimental situation this way of reasoning amounts to saying almost nothing. First, there is no measurable property of a single system that could be interpreted as a probability. Second, and this is more important, a pure state φ provides already maximal information about the system $S(\varphi)$. Indeed, the values of all objective properties can be deduced from the knowledge of φ, and these values will be obtained with certainty by a measuring process that fulfils the calibration postulate.

Moreover, by means of probability theorem I it follows from the general validity of the calibration postulate that the numbers $p(\varphi, a_i)$ are values of a relative frequency observable F_k^N that pertains to an ensemble $\{Z_{l_i}\}^N$ of N pointer values Z_{l_i} obtained by A-measurements on N equally prepared systems $S_n(\varphi)$, in the limit $N \to \infty$. In other words, on the basis of the probability-free interpretation I_M^0 the values $p(\varphi, a_i)$ turn out to be objective properties of the ensemble $\{Z_{l_i}\}^N$. These arguments show that for an individual system $S(\varphi)$ there is no need for an additional 'potentia' or 'propensity', since from the state φ one can deduce all objective properties, even the statistical ones that turn out to be yes–no properties of a conveniently defined ensemble. For a single system $S(\varphi)$, the totality of probabilities $p(\varphi, a_i)$, $p(\varphi, b_k)$ expresses nothing but the fact that φ is the state of S.

In the case of repeatable measurements, one can apply also probability theorem II. Under the assumption that the calibration postulate holds for a single system $S(\varphi)$ as well as for a compound system $S^{(N)}$, one arrives at the following result. The numbers $p(\varphi, a_i)$ are not only the relative frequencies of pointer values $Z_i = f^{-1}(a_i)$ in the ensemble of pointer outcomes but also the relative frequencies of systems $S_i(\varphi^{a_{l_i}})$ in the ensemble $S^{(N)}$ of originally equally prepared systems after A-measurements. This means that the relative frequency observable f_k^N pertains to the post-measurement compound system $S^{(N)}$ as an objective observable, the value of which is given by $p(\varphi, a_i)$, at least in the limit $N \to \infty$.

Summarizing the preceding arguments, we arrive at the following result. A single quantum system $S(\varphi)$ with preparation φ possesses values a_i of all objective observables $A = \sum a_i P(a_i)$. According to the calibration postulate, these values can be measured with certainty. For a nonobjective observable B, nothing is known about its values b_k, and no prediction can be made for the result of a B-measurement. The probabilities $p(\varphi, b_k)$ that are induced by φ and B do not refer to the system $S(\varphi)$ but to an ensemble $\{Z_{l_i}\}^N$ of pointer values or to an ensemble $\{S_i(\varphi^{b_{l_i}})\}^N$ of N equally prepared systems $S_i(\varphi)$ after measurements of B. According to probability theorems I or II, the values $p(\varphi, b_k)$ are eigenvalues of the relative frequency observables of the ensembles $\{Z_{l_i}\}^N$ or $S^{(N)}$ respectively, in the limit of infinite N.

Obviously, the probability-free interpretations I_M^0 and I_R^0 are completely sufficient for the interpretation of a quantum system. Any interpretation of an individual system that goes beyond the calibration postulates and makes use of 'metaphysical' concepts like *potentiality* or *propensity* is purely hypothetical and no longer justified by quantum mechanics. There is no need and no room for further interpretations of the theoretical terms $p(\varphi, a_i)$. For an individual system S with preparation φ the values $p(\varphi, a_i)$ give no additional information except that S is in the state φ. Hence one gets the impression that the problem of the single-case interpretation of quantum mechanical probabilities is actually a pseudo-problem.

We conclude this section with some further historical remarks. Since the discovery of quantum mechanics in the 1920s, this theory has been considered as intimately related to the concepts of uncertainty, indeterminacy, and probability. Within the framework of different interpretations, these concepts have been understood as subjective ignorance, as deficiency of information, as objective indeterminacy, as potentiality, or as propensity. But without referring to any of these meanings, the statistical content of the concepts has been incorporated into the 'minimal interpretation' of quantum mechanics.

In contrast to this well-established understanding of quantum theory, probability theorems I and II have led to the surprising result that quantum mechanics is primarily not a probabilistic theory but a theory that is concerned with yes–no propositions. A result of this kind was in fact conjectured by several authors, e.g. Everett [Eve 57], Finkelstein [Fink 62], Hartle [Hart 68], Graham [Gra 73], and DeWitt [DeW 71]. However, instead of referring to the reduced state of the object system or of the apparatus after the measurement, these authors prove an analogue of probability theorem II for a pure-state preparation prior to the measurement (probability theorems III and IV). The present investigations have shown that the original conjectures were nevertheless correct and can be justified within the quantum theory of measurement by means of probability theorems I and II.

4

The problem of objectification

4.1 The concept of objectification

4.1(a) Weak and strong objectification

Let S be a quantum system that is prepared in a pure state φ and A a discrete nondegenerate observable with eigenvalues a_i and eigenstates φ^{a_i}.

If the preparation φ is not an eigenstate of A, then quantum mechanics does not provide any information about the value of the observable A. The pair $\langle \varphi, A \rangle$ merely defines a probability distribution $p(\varphi, a_i)$, the experimental meaning of which is given by the statistical interpretation of quantum mechanics: the real positive number $p(\varphi, a_i)$ is the probability of obtaining the result a_i after a measurement process of the observable A. This means that if one were to perform a large series of N measurements of this kind, then the relative frequency $f^N(\varphi, a_i)$ of the result a_i would approach for almost all test series the probability $p(\varphi, a_i)$, for $N \to \infty$ (cf. chapter 3). One could, however, in addition to these well-established results tentatively assume that a certain value a_i or even an eigenstate φ^{a_i} of A pertains objectively to the system S but that this value or state is subjectively unknown to the observer, who knows only the probability $p(\varphi, a_i)$ and hence the distribution of possible measurement outcomes. The probability would then express the subjective ignorance (or knowledge) about an objectively decided value or eigenstate of A. The hypothetical attribution of a certain value or eigenstate of A to the system S will be called *objectification*.

Here we distinguish three kinds of objectification assumptions, which are conceptually of different strength and of different explanatory power. The strongest hypothesis is that the system is actually in an eigenstate φ^{a_i} of A. In this case it follows that S possesses also the corresponding eigenvalue a_i of A. The state φ would then merely describe the observer's incomplete knowledge about the actual state φ^{a_i} of the system. The hypothetical attribution of a certain eigenvalue of A to the system S will be called strong objectification. We summarize this assumption as:

The strong objectification hypothesis (state-attribution). It is possible to attribute an eigenstate φ^{a_i} of A to the system S in state φ such that S is actually in φ^{a_i} but this state is unknown to the observer.

Strong objectification can be relaxed by assuming that the system merely possesses objectively a value a_i of the observable A, i.e., that a value a_i can be attributed to the system in a consistent way. It is obvious that from this hypothesis it does not necessarily follow that an eigenstate φ^{a_i} pertains to the system in question. The hypothetical attribution of a certain value of A to the system S will be called weak objectification. We summarize this assumption as:

The weak objectification hypothesis (value-attribution). It is possible to assign a value a_i of the observable A to the system S in the state φ such that the value a_i pertains objectively to the system but this value is subjectively unknown to the observer.

One could try to relax this weak objectification hypothesis further. This can be done in the following way. If weak objectification of an observable A is assumed, then it follows that to each of its objectified values a_i a probability $p(\varphi, a_i)$ is attached, which characterizes the probability distribution of the values a_i and which is reproduced in the statistics of A-measurements. Hence, weak objectification of the values a_i implies that to the system S with preparation φ probabilities $p(\varphi, a_i)$ are also attributed as hypothetical entities. Since the converse is not generally true, one could dispense with value-attribution but nevertheless preserve probability-attribution. In this way, we arrive at the weakest objectification hypothesis that we shall consider:

Very weak objectification (probability-attribution). For a system S with preparation φ one can attribute probabilities $p(\varphi, a_i)$ of A-values a_i in a consistent way. The values $p(\varphi, a_i)$ refer to the system S and are subject to the well-known probability axioms (e.g. Kolmogorov).

From a conceptual point of view the three assumptions are of different strength and different explanatory power. State-attribution is stronger than value-attribution, and value-attribution is stronger than probability-attribution. The three hypotheses are attempts to go beyond the realistic interpretation and to attach new entities to the object system. However, it must be clarified whether the observable (empirical) consequences of the hypothetical assumption are really of different strengths. This can only be ascertained by a detailed investigation. A comparison between the concepts of weak and strong objectification will be carried through in 4.2 and 4.3, whilst the problem of probability attribution will be considered in 4.4.

4.1(b) Pure and mixed states

On the one hand, the concept of objectification refers to entities such as eigenstates, eigenvalues, and probabilities, which are attributed to a quantum system. On the other hand, a quantum system S is characterized by its state. Hence, the possibilities of objectification will depend not only on the objectified entity but also on the state of the quantum system in question. Here we consider pure and mixed states; moreover, two kinds of mixed state must be distinguished.

If a proper quantum system S with Hilbert space H_S is prepared in a *pure state* $\varphi \in H_S$, then the observer possesses maximal information about S. The property P_φ given by $P[\varphi]$ pertains to the system and the totality of properties P_i given by the projection operators P_i that fulfil the commutator relation $[P_\varphi, P_i] = 0$ are objective with respect to the system S. This means that for each property P_i either P_i or its contrary $\neg P_i$ pertains to the system. The preparation of a pure state can be performed by, e.g., repeatable measurements of a discrete nondegenerate observable.

On the other hand, a quantum system can also be prepared in a mixed state W, where $W = W^+$ is a self-adjoint operator with $\operatorname{tr} W = 1$. As already mentioned, there are two ways to prepare a mixed state, and they lead to different results. The first, less interesting, way makes use of a preparation apparatus, e.g. an accelerator, that does not work completely accurately and prepares systems in states φ_i, say, with a priori probabilities p_i which depend on the construction of the apparatus and which can be determined separately. In this case, any single system S is actually in one of the states φ_i; which one, however, is not known to the observer, who knows only the probabilities p_i. This special kind of mixed state will be called a mixture of states or Gemenge, as mentioned earlier. A Gemenge $\Gamma(p_i, \varphi_i)$ is a classical mixture of states φ_i with weights p_i without any quantum mechanical peculiarities. Nevertheless, the system can be described also by the mixed state $W = W(\Gamma) = \sum p_i P[\varphi_i]$, since this mixed state leads to the same statistical predictions as the Gemenge $\Gamma(p_i, \varphi_i)$. Irrespective of the fact that the decomposition of W into orthogonal or even nonorthogonal states is by no means unique, one can obviously apply the ignorance interpretation to this state. Since the state W was prepared as a mixture of states φ_i, the assumption that the system S described by state W is actually in a certain state φ_i cannot lead to any difficulties.

The second way of preparing a quantum system in a mixed state is by separation. Let $S = S_1 + S_2$ be a compound system, with Hilbert space H, which consists of two subsystems S_1 and S_2 with Hilbert spaces H_1 and H_2 such that $H = H_1 \otimes H_2$. If S is prepared in a pure state $\Phi(S)$ then the subsystems S_1 and S_2 are in the reduced mixed states $W(S_1) = \mathrm{tr}_2 P[\Phi]$ and $W(S_2) = \mathrm{tr}_1 P[\Phi]$, where tr_k, $k = 1, 2$, denotes the partial trace with respect to system S_k. To say that the subsystem S_1 is in a mixed state $W(S_1)$ means that one considers only those properties of the total system S that are concerned with the degrees of freedom of system S_1, neglecting in this way all possible correlations between S_1 and S_2. Because the correlations of this kind are neglected, the description of S_1 is no longer maximal, which means that $W(S_1)$ is a mixed state. Hence, if the system S_1 can be separated (in space) from S_2 without changing the state Φ of the compound system, the system S_1 will in general be in a mixed state $W(S_1)$. This state is a genuine mixed state, and the question whether it admits an 'ignorance interpretation' is more complicated and requires a separate investigation.

Summarizing these arguments, we find that there are four objectification problems. The question of the extent to which the strong or the weak objectification hypothesis is compatible with quantum mechanics will be discussed for the two cases in which the state of the system considered is either a pure state or a mixture (see Table 4.1).

Table 4.1.

	Strong objectification?	Weak objectification?
pure state	I	II
mixed state	III	IV

4.2 Objectification in pure states

Let us consider first the simplest but nevertheless also the most important case, of a proper quantum system S (without super-selection rules) with Hilbert space H_S, in a pure state $W = P[\varphi]$, $\varphi \in H_S$. Let A be a discrete observable with spectral decomposition $A = \sum_i a_i P[\varphi^{a_i}]$, where the a_i are the eigenvalues of A and the $P[\varphi^{a_i}]$ are the projection operators that project onto the eigenstates φ^{a_i} of A. In the more general case of degenerate eigenvalues, we write $A = \sum a_i P(a_i)$, where $P(a_i)$ projects onto the subspace of value a_i. By A_i we denote the yes–no proposition that the value a_i of A pertains to the system, or the property of a system of possessing the value a_i. Furthermore, we assume that the state φ is not an eigenstate of A, i.e., $[P[\varphi], P(a_i)] \neq 0$ for all eigenvalues a_i. The pair $\langle W, A \rangle$ then defines a probability distribution $p(\varphi, A_i)$ with

$$p(\varphi, A_i) = \text{tr}\{P[\varphi] P(a_i)\}.$$

According to the probability interpretation of quantum mechanics, the number $p(\varphi, A_i) \in [0, 1]$ is the probability of obtaining the result a_i if the observable A is measured on the system S with preparation φ (cf. chapter 3).

4.2(a) Strong objectification hypothesis

The assumption that the observable $A = \sum a_i P(a_i)$ can be strongly objectified with respect to the system S in the state $W = P[\varphi]$ means that S is actually in an eigenstate W^{a_i} of A (with value a_i) but that this state is unknown to the observer, who knows only the probabilities $p_i = p(\varphi, A_i)$. Hence the system is in a 'mixture of states' $\{p_i, W^{a_i}\}$, which can be described by the *mixed state*

$$\tilde{W} = \sum p(\varphi, A_i) W^{a_i}. \tag{4.1}$$

Since the W^{a_i} are eigenstates of A, it follows that $A W^{a_i} = a_i W^{a_i}$, and for the state $\tilde{W}$ one obtains

$$\tilde{W} = \sum P(a_i) \tilde{W} P(a_i). \tag{4.2}$$

Because the right-hand side of this equation is the Lüders mixture $W_L(\tilde{W}, A)$ that one would obtain by a Lüders measurement of A on the system S in the state $\tilde{W}$, eq. (4.2) is not changed by a measurement (without reading) of this kind.

Let $B = \sum_k b_k P[\psi^{b_k}]$ be another observable (or more generally $B = \sum_k b_k P(b_k)$) such that the two relations

$$[A, B] \neq 0, \qquad [W, B] \neq 0$$

are fulfilled. If one assumes that S is in the pure state $P[\varphi]$ and that the observable A can be strongly objectified, then on the basis of (4.2) one obtains for the probability $p(\varphi, B_k) = \text{tr}\{P[\varphi] P(b_k)\}$ of the value b_k of B

$$p(\varphi, B_k) = \text{tr}\left\{\sum_i P(a_i)P[\varphi]P(a_i)P(b_k)\right\}$$

$$= \text{tr}\{W_L(\varphi, A) P(b_k)\} \tag{4.3}$$

where $W_L(\varphi, A) = \sum P(a_i)P[\varphi]P(a_i)$ is the Lüders mixture that one obtains after a Lüders measurement (without reading) of A performed on the system S in state φ. Equation (4.3) is the formal expression of strong objectification of the observable A with respect to $S = S(\varphi)$, tested by means of the special observable $P(b_k)$. On the one hand, clearly strong objectification of A implies that the condition (4.3) holds for arbitrary test-observables P_T. On the other hand, for the probability $p(\varphi, B_k) = \text{tr}\{P[\varphi]\,P(b_k)\}$ one obtains by an elementary quantum mechanical calculation

$$p(\varphi, B_k) = \text{tr}\{W_L(\varphi, A)\,P(b_k)\} + p_{\text{int}}(\varphi; B_k, A) \qquad (4.4)$$

with an 'interference term'

$$p_{\text{int}}(\varphi; B_k, A) = \text{tr}\left\{\sum_{i \neq j} P[\varphi]P(a_j)P(b_k)\right\}.$$

Therefore, the assumption of strong objectification applied to A requires that the interference term vanishes for arbitrary test observables B, i.e.,

$$\forall B, \qquad p(\varphi; B_k, A) = 0. \qquad (4.5)$$

However, this is known to be true only for eigenstates φ of A. In all other cases there are some observables B such that $p_{\text{int}}(\varphi; B_k, A) \neq 0$ and (4.5) is violated. Hence, for arbitrary preparation states φ the hypothesis of strong A-objectification is not consistent with quantum mechanics.

4.2(b) Illustration by an experiment

The result that strong objectification in a pure state is in general not possible can be illustrated by the photon split beam experiment shown in Fig. 4.1. In the experimental setup, a Mach–Zehnder interferometer, the state φ of the incoming photon is split by a semi-transparent mirror – the beam splitter BS_1 – into two components described by orthonormal states ψ^B and $\psi^{\neg B}$. The two parts of the split beam are reflected at two fully reflecting mirrors M_1 and M_2 and recombined with a phase shift δ at a second semi transparent mirror – the beam splitter BS_2. In this experiment there are two mutually exclusive measuring arrangements. If the photon detectors D_1 and D_2 are in the position $(D_1{}^B, D_2{}^B)$, one observes which way (B or $\neg B$) the photon came. If the detectors are in the position $(D_1{}^A, D_2{}^A)$ one observes an interference pattern, i.e., intensities that depend on the phase difference δ. Hence, the observables of the path and of the interference pattern are complementary in the sense of Niels Bohr.

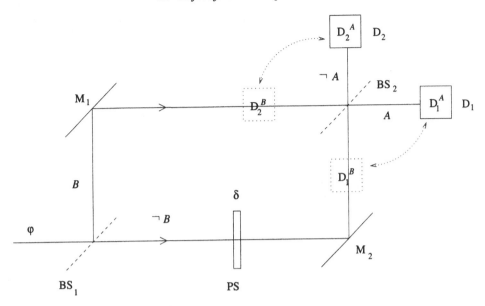

Fig. 4.1 Photon split beam experiment with beam splitters BS_1 and BS_2 (both semireflecting mirrors), two fully reflecting mirrors M_1 and M_2, a phase shifter PS and two photon detectors D_1 and D_2 in positions (D_1^B, D_2^B) or (D_1^A, D_2^A).

The photon split beam experiment can be completely described within the framework of a two-dimensional Hilbert space H_2. The observables are then given by projection operators $P(B)$ (for the path) with eigenstates ψ^B and $\psi^{\neg B}$ and $P(A)$ (for the interference pattern), with eigenstates φ^A and $\varphi^{\neg A}$. The preparation state φ of the incoming photon can be decomposed in the orthonormal basis $\{\psi^B, \psi^{\neg B}\}$ as

$$\varphi = \tfrac{1}{\sqrt{2}}(\psi^B + e^{i\delta}\,\psi^{\neg B}), \tag{4.6}$$

where $\delta \in R^+$ is the phase shift. Hence, φ is not an eigenstate of $P(B)$ and the probabilities for the ways B and $\neg B$ are given by

$$p(\varphi, B) = \tfrac{1}{2}, \qquad p(\varphi, \neg B) = \tfrac{1}{2}, \tag{4.7}$$

respectively. The interference observable, which can be measured by detector D_1, is given by $P(A)$ with eigenstates

$$\varphi^A = \tfrac{1}{\sqrt{2}}(\psi^B + \psi^{\neg B}), \quad \varphi^{\neg A} = \tfrac{1}{\sqrt{2}}(\psi^B - \psi^{\neg B}).$$

Hence, the probabilities for A (the photon is registered in D_1) and for $\neg A$ (the photon is registered in D_2) are given by

$$p(\varphi, A) = \cos^2(\delta/2), \qquad p(\varphi, \neg A) = \sin^2(\delta/2), \tag{4.8}$$

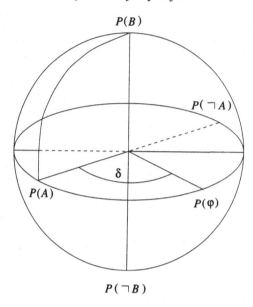

Fig. 4.2 Poincaré sphere of the photon split beam experiment shown in Fig. 4.1.

respectively. The complementarity between the path and the interference observable is formally expressed by the noncommutativity of $P(B)$ and $P(A)$, i.e., by $[P(b), P(A)] \neq 0$. States and operators of this two-dimensional Hilbert space can be illustrated by means of the Poincaré sphere $\mathscr{P}$ (cf. 2.4(c)). If the axes of $\mathscr{P}$ are chosen such that the path operator $P(B)$ reads

$$P(B) = \tfrac{1}{2}(\sigma_3 + 1)$$

then the operator $P(A)$ of the interference pattern is given by

$$P(A) = \tfrac{1}{2}(\sigma_1 + 1),$$

and $P(A)$ and $P(B)$ correspond to orthonormal vectors $\mathbf{A}$ and $\mathbf{B}$. In the Poincaré-sphere picture the orthogonality of two vectors represents the complementarity of the respective observables.

The initial preparation $P[\varphi]$ with $\varphi = \frac{1}{\sqrt{2}}(\psi^B + e^{i\delta}\psi^{\neg B})$ is given here by a unit vector $\mathbf{n}_\varphi$, or a point on the equator of the Poincaré sphere. The phase shift δ corresponds to the angle between the unit vectors $\mathbf{n}_\varphi$ and $\mathbf{A}$ in the horizontal plane. (Fig. 4.2)

The strong objectification hypothesis can now be discussed within the framework of the photon split beam experiment. The preparation φ is not an eigenstate of the path observable $P(B)$, and the observables $P(B)$ and $P(A)$ do not commute. If one assumes that the photon is actually in the state $\psi^{\neg B}$ (with probability $\tfrac{1}{2}$), then one is assuming that $P(B)$ can be strongly

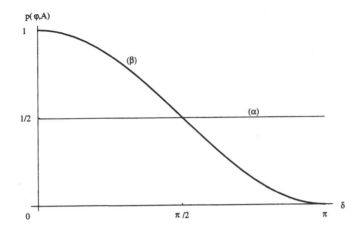

Fig. 4.3 The intensity $I_1 = p(\varphi, A)$ in the detector D_1 (α) according to the objectification hypothesis; and (β) according to quantum mechanics.

objectified. This means that the photon can be described by the mixed state

$$W_L(\varphi, B) = \tfrac{1}{2}P(B) + \tfrac{1}{2}P(\neg B) \qquad (4.9)$$

i.e., by the Lüders mixture that one would obtain after a Lüders measurement of $P(B)$ on the photon in state φ.

In the Poincaré sphere, this mixed state $W_L = W(\mathbf{Y})$ corresponds to the centre of the sphere, i.e., to a null vector $\mathbf{Y} = 0$. Using the mixed state $W_L(\varphi, B)$, for the probability $\tilde{p}(\varphi, A)$ of registering the photon in the detector D_1 one obtains in accordance with (4.3)

$$\tilde{p}(\varphi, A) = \mathrm{tr}\{W_L(\varphi, B)P(A)\} = \tfrac{1}{2} \qquad (4.10)$$

and also $\tilde{p}(\varphi, \neg A) = \tfrac{1}{2}$. It should be noted that, on the one hand, these probability values do not depend on the phase shift δ. On the other hand the quantum mechanical calulation of $p(\varphi, A)$, say, yields

$$p(\varphi, A) = \mathrm{tr}\{P[\varphi]P(A)\} = \cos^2(\delta/2) \qquad (4.11)$$

in disagreement with (4.10), which follows from the strong objectification hypothesis. This means that the interference term $p_{\mathrm{int}}(\varphi; B, A)$ given by

$$\underbrace{p(\varphi, A)}_{\cos^2(\delta/2)} = \underbrace{\tilde{p}(\varphi, A)}_{\tfrac{1}{2}} + \underbrace{p_{\mathrm{int}}(\varphi; B, A)}_{\tfrac{1}{2}\cos\delta} \qquad (4.12)$$

reads $p_{\mathrm{int}}(\varphi; B, A) = \tfrac{1}{2}\cos\delta$ (Fig. 4.3).

The quantum mechanical probabilities $p(\varphi, A)$ and $p(\varphi, \neg A)$, which correspond to the intensities I_1 and I_2 measured by means of the detectors D_1

and D_2, respectively, show the interference pattern under consideration. The measured intensities I_1 and I_2 depend on the phase shift δ in accordance with (4.12). On the other hand, the intensities that follow from the strong objectification hypothesis do not depend on δ at all. Hence, a comparison of the two probabilities $p(\varphi, A)$ and $\tilde{p}(\varphi, A)$ – as illustrated in Fig. 4.3 – shows that the path observable $P(B)$ cannot be objectified whenever $p_{\text{int}}(\varphi; B, A) \neq 0$.

4.2(c) Weak objectification

Since the strong objectification hypothesis of an observable $A = \sum a_i P(a_i)$ with respect to an arbitrary pure state φ is not compatible with quantum mechanics, one could try to relax this hypothesis in the sense of weak objectification. This means that at least a value a_i of A can be attributed to a system S with preparation φ. Let $B = \sum b_k P(b_k)$ be another observable with values b_k. The propositions A_i and B_k, stating that the values a_i and b_k, respectively, pertain to the system S, will then fulfil the logical equivalence

$$B_k = (A_i \wedge B_k) \vee (\neg A_i \wedge B_k) \tag{4.13}$$

which is known to be valid in classical logic. This logic can be formally expressed by a Boolean lattice L_B. By $\neg A_i$ we denote the negation of A_i, which corresponds to the projection operator $P(\neg a_i) = 1 - P(a_i)$ that maps onto the completely orthonormal subspace $M(a_i)$. The meaning of the equivalence (4.13) is obvious: if b_k pertains to the system, then, for another observable A, either a_i or $\neg a_i$ (i.e., any other value $a_j, j \neq i$) pertains to the system, which means that A is objective. On the basis of the Boolean lattice L_B, probabilities can be introduced by a measure

$$p : \quad L_B \longrightarrow [0, 1],$$

which is a mapping from the Boolean lattice of propositions into [0,1] that satisfies the

Kolmogorov axioms

(i) $p(0) = 0, \qquad p(\mathbf{I}) = 1$

(ii) $p(\neg a) = 1 - p(a)$ *for all* $a \in L_B$ $\qquad\qquad$ (4.40)

(iii) *for any sequence* $\{A_i\}, A_i \in L_B, A_i \leq \neg A_k,$
$\quad i \neq k,$ *the relation* $p(A_1 \vee A_2 \vee \cdots) = p(A_1) + p(A_2) + \cdots$ *holds.*

From the logical equivalence (4.13), one obtains by means of this probability measure the probabilistic equation

$$p(\varphi, B_k) = p(\varphi, A_i \wedge B_k) + p(\varphi, \neg A_i \wedge B_k). \tag{4.14}$$

Instead of considering only a pair $(A_i, \neg A_i)$ of alternative propositions, one can also consider the totality $\{A_1, A_2, \ldots\}$ of alternative propositions A_i $(A_i \leq \neg A_k, i \neq k)$ that correspond to the eigenvalues $\{a_1, a_2, \ldots\}$ of A. In this case, (4.14) must be replaced by the more general equation

$$p(\varphi, B_k) = \sum_i p(\varphi, A_i \wedge B_k). \tag{4.15}$$

The probability $p(\varphi, A_i \wedge B_k)$ is the probability of obtaining both the value a_i and the value b_k, i.e., the joint probability for a_i and b_k. Obviously, on account of the incommensurability of A and B it is in general not possible to give this probability a direct experimental meaning. Instead, one can determine experimentally only the probability $p(\varphi, A_i \sqcap B_k)$ for the sequential proposition $A_i \sqcap B_k$, as follows. If after a measurement of A with outcome a_i another observable B is measured with the result b_k, then the corresponding sequential proposition $A_i \sqcap B_k$ means that first a_i and then b_k is found to pertain to the system. In the framework of quantum measurements, this proposition is defined with reference to proof processes that are given by Lüders measurements. Accordingly, the sequential probability $p(\varphi, A_i \sqcap B_k)$ is given by

$$p(\varphi, A_i \sqcap B_k) = \mathrm{tr}\{P(a_i)P[\varphi]P(a_i)P(b_k)\}.$$

If one compares this term with the well-known expression for the joint probability in Hilbert space,

$$p(\varphi, A_i \wedge B_k) = \lim_{n \to \infty} (\varphi, \{P(a_i)P(b_k)\}^n \varphi)$$

one obtains the relation

$$p(\varphi, A_i \wedge B_k) \leq p(\varphi, A_i \sqcap B_k). \tag{4.16}$$

This inequality holds, since for $A_i \wedge B_k$ it is not required that after the B-measurement the A-result a_i still pertains to the system. From (4.15) and (4.16) one obtains the relation

$$p(\varphi, B_k) \leq \sum_i p(\varphi, A_i \sqcap B_k) \tag{4.17}$$

which holds for the observable probabilities $p(\varphi, B_k)$ and $p(\varphi, A_i \sqcap B_k)$. Using the definition

$$W_L(\varphi, A) = \sum_i P(a_i)P[\varphi]P(a_i) \qquad (4.18)$$

for the Lüders mixture, the right-hand side of (4.17) reads

$$\sum_i p(\varphi, A_i \sqcap B_k) = \text{tr}\{W_L(\varphi, A)P(b_k)\}. \qquad (4.19)$$

Hence, one obtains

$$p(\varphi, B_k) \leq \text{tr}\{W_L(\varphi, A)P(b_k)\} \qquad (4.20)$$

This inequality is the formal expression for the weak objectification of A with respect to the state φ, tested by means of the observable $P(b_k)$.

If one compares the weak objectification condition (4.16) with eq. (4.3) for strong objectification, one gets the impression that weak objectification is indeed a relaxation of strong objectification. Whereas strong objectification (4.3) is violated by the quantum mechanical equation (4.4) whenever $p_{\text{int}}(\varphi; A, B_k) \neq 0$, the condition (4.16) for weak objectification is violated by quantum mechanics only if $p_{\text{int}}(\varphi; A, B_k) > 0$. However, this way of reasoning is not quite correct. Weak objectification of A with respect to φ implies that the inequality (4.20) must hold for arbitrary test observables B and all values of k. If one replaces $P(b_k)$ by $P(\neg b_k) = 1 - P(b_k)$ and the probability $p(\varphi, B_k)$ by $p(\varphi, \neg B_k) = 1 - p(\varphi, B_k)$, one obtains the new condition

$$p(\varphi, B_k) \geq \text{tr}\{W_L(\varphi, A)P(b_k)\} \qquad (4.21)$$

Taking together conditions (4.20) and (4.21) one arrives at the equation

$$p(\varphi, B_k) = \text{tr}\{W_L(\varphi, A)P(b_k)\}$$

which is equivalent to the strong objectification condition (4.3).

These arguments show that the hypothesis of weak objectification, or value-attribution, leads finally to the same equation as that of strong objectification (state-attribution). Since the weak objectification condition (4.20) must hold for arbitrary test-observables B, the inequality (4.20) can always be strengthened into the equality (4.3). Since value-attribution is conceptually weaker than state-attribution, one should start in general from the weak objectification hypothesis. However, since this assumption leads to the same equation as strong objectification, it is incompatible with quantum mechanics as well. Hence, we find that the weak objectification of an observable A with respect to an arbitrary pure state φ is also not consistent with quantum mechanics.

4.2(d) Quantum-logical resolution

The hypothesis of weak objectification consists not only of attributing values a_i or properties A_i to a system S in a pure state φ. In addition, one makes use of the assumption that the properties A_i fulfil the laws of classical logic. In particular, for two different properties A_i and B_k the logical equivalence (4.13),

$$B_k = (A_i \wedge B_k) \vee (\neg A_i \wedge B_k),$$

must be presupposed. This equivalence holds in classical logic, which can be formally expressed by a Boolean lattice L_B. Probabilities can then be introduced by a measure

$$p : \quad L_B \longrightarrow [0,1],$$

which is a mapping of the Boolean lattice L_B of propositions into $[0,1]$ satisfying the Kolmogorov axioms given above. From the logical equivalence (4.13), one obtains by means of this probability measure the probabilistic equation (4.14),

$$p(\varphi, B_k) = p(\varphi, A_i \wedge B_k) + p(\varphi, \neg A_i \wedge B_k),$$

which turns out to be inconsistent with quantum mechanics.

Instead of refuting the weak objectification hypothesis, one can also query the validity of classical logic for the domain of quantum mechanical propositions. Instead, for quantum mechanical propositions one can justify only the weaker *quantum logic* corresponding to an orthomodular lattice L_q that is atomic and fulfils the covering law. This justification results from the most general conditions under which quantum mechanical propositions can be tested in accordance with the restrictions of the quantum theory of measurement. Within the framework of the orthonormal lattice L_q the equivalence (4.13) is no longer generally valid unless the propositions B_k and A_i are known to be commensurable. Instead, in the lattice L_q only the weaker implication relation

$$(A_i \wedge B_k) \vee (\neg A_i \wedge B_k) \leq B_k \qquad (4.22)$$

holds for arbitrary elements A_i and B_k. Let us now consider the non-Boolean probability measure

$$p : \quad L_q \longrightarrow [0,1]$$

mapping the lattice L_q into $[0,1]$, where the probability function $p(\varphi, A)$, $A \in L_q$, again satisfies the probability axioms mentioned above. From the quantum logical implication (4.22), one obtains by means of this probability measure the inequality

$$p(\varphi, B_k) \geq \sum_i p(\varphi, A_i \wedge B_k), \tag{4.23}$$

where the sum runs over all properties A_i. Obviously, (4.23) is weaker than (4.15). This means that weak objectification of an observable A, i.e., the attribution of A-values to a quantum system, leads to the probabilistic relation (4.23) if one presupposes that the properties A_i fulfil only the laws of ortho-modular quantum logic. Hence, the question arises whether this *quantum-logical weak objectification* is still incompatible with quantum mechanics, in particular with the quantum mechanical interference phenomenon.

The quantum mechanical equation (4.4) can be reformulated in terms of sequential probabilities, making use of (4.19):

$$p(\varphi, B_k) = \sum_i p(\varphi, A_i \sqcap B_k) + p_{\text{int}}(\varphi; B_k, A). \tag{4.24}$$

Since (4.16) holds in classical as well as in quantum probability, we obtain the quantum mechanical inequality

$$p(\varphi, B_k) \geq \sum_i p(\varphi, A_i \wedge B_k) + p_{\text{int}}(\varphi; B_k, A). \tag{4.25}$$

If one compares the quantum mechanical relations (4.24) and (4.25) with the quantum-logical weak objectification conditions (4.23), one finds that these relations are fully consistent. Indeed, (4.23) can also be derived from Hilbert-space quantum mechanics. For arbitrary projection operators $P(b)$ and $P(a)$ the relation $P(b) \geq P(a \wedge b) + P(\neg a \wedge b)$ holds, and hence one obtains for the probabilities $p(\varphi, b) = (\varphi, P(b)\varphi)$ etc.

$$p(\varphi, b) \geq p(\varphi, a \wedge b) + p(\varphi, \neg a \wedge b).$$

An obvious generalization of this inequality leads to (4.23). Hence, the quantum mechanical equation (4.24) cannot contradict the condition (4.23), which is generally valid in quantum mechanics. (Clearly, the consistency of (4.23) with (4.24) and (4.25) can also be shown by explicit calculation.) This means that quantum-logical weak objectification is compatible with the quantum mechanical interference phenomenon. In the photon split beam experiment mentioned above (section 4.2(b)), the quantum-logical weak objectification (4.23) is almost trivial and reads $p(\varphi, A) \geq 0$. Clearly, this condition is in complete accordance with the quantum mechanical equation (4.24). This can also easily be seen in Fig. 4.3.

However, it should be emphasized that this resolution of the contradiction between the weak objectification hypothesis and quantum probability does

not mean that a nonobjective property can be objectified in the original
sense. In classical language and logic one can express the objectification
in such a way that it contradicts quantum probability. However, the more
restrictive quantum language and logic generally do not allow one to make
statements that are inconsistent with quantum probability. The reason is
that the various quantum mechanical restrictions that derive from the mutual incommensurability of quantum mechanical propositions are already
incorporated into the formal logic of quantum language. In particular, this
means that the value-attribution hypothesis expressed in terms of quantum
logic leads to the quantum-logical weak objectification, which is compatible
with quantum mechanics. Needless to say, that this relaxed quantum-logical weak objectification requirement no longer expresses the possibility of
attributing values to the considered system in the sense of classical logic.

These considerations show, in particular, that weak objectification, or
value-attribution, must be refuted in quantum mechanics only if, in addition
to the attribution of values a_i to the object system, the validity of classical
logic for the propositions A_i is assumed. Usually this presupposition is considered to be almost trivial. However, the investigation of the last subsection
has shown that the quantum-logical relaxation of the laws of logic would
allow for a quantum-logical weak objectification. Hence, we emphasize that
value-attribution together with classical propositional logic is untenable in
quantum mechanics.

4.3 Objectification in mixed states

4.3(a) Mixed states versus mixtures of states

From a formal point of view, a mixed state W of a proper quantum system S
with Hilbert space $\mathcal{H}_S$ is given by a positive self-adjoint operator $W = W^+$
with tr $W = 1$. A mixed state of this kind must be clearly distinguished
from a 'mixture of states', i.e., an ensemble $\Gamma(p_i, \varphi_i)$ of pure states $\varphi_i \in \mathcal{H}$
and statistical weights (probabilities) p_i. There are at least two situations in
which a system S cannot be described by a pure state – which would provide
maximal information – but only by a mixed state:

1. The pure state φ_i of the system S is not known to the observer but
   only the probabilities p_i.
2. The system S is a subsystem of a compound system $\hat{S} = S + S'$, where
   $\hat{S}$ is in a pure state $\hat{\Psi}(S + S')$.

It is obvious that in the first case there is no essential difference between the Gemenge $\Gamma(p_i, \varphi_i)$ and the mixed state $W = \sum p_i P[\varphi]$, since the latter may be considered merely as the description of the Gemenge $\Gamma(p_i, \varphi_i)$ in terms of Hilbert-space quantum mechanics. This means that there are obviously no difficulties in interpreting the mixed state $W = \sum p_i P[\varphi]$ as the description of a system that possesses one of the states φ_i with probability p_i. In this situation, we say that the state W admits an 'ignorance interpretation'.

In the second case, the situation is much more difficult. If a compound system $S = S_1 + S_2$ with Hilbert space $\mathscr{H}_S = \mathscr{H}_1 \otimes \mathscr{H}_2$ is prepared in a pure state $\Phi(S) \in \mathscr{H}_S$, the mixed state of the subsystem S_1, say, is given by the partial trace $W(S_1) = \mathrm{tr}_2 P[\Phi]$, which is a summation over the degrees of freedom of system S_2. It is easy to show that $W(S_1)$ is a positive self-adjoint operator with $\mathrm{tr}\, W(S_1) = 1$, i.e., a mixed state in the formal sense. However, nothing is known about the decomposition of this mixed state into weighted components corresponding to pure states. There are two arguments that show why a mixed state W does not determine the mixture of states $\Gamma(W)$ that is formally described by it.

(i) If the spectrum of the self-adjoint operator W is not degenerate, then there exists a spectral decomposition

$$W = \sum_{i=1}^{\infty} p_i P[\psi_i]$$

of W into mutually orthogonal, i.e., mutually exclusive pure states ψ_i. The states ψ_i are the eigenstates of W and the coefficients p_i the corresponding eigenvalues. Hence, the eigenvalue equation $W\psi_i = p_i\psi_i$ holds for any $i \in \mathbb{N}$. From the positivity of the operator W, it follows that $p_i \geq 0$ and from $\mathrm{tr}\, W = 1$ the normalization condition $\sum_{i=1}^{\infty} p_i = 1$ for the coefficients p_i follows. Hence, one could (tentatively) consider the real positive numbers p_i as the probabilities for the states ψ_i. The state W would then be a formal description of the Gemenge $\Gamma(p_i, \psi_i)$.

However, the decomposition of the state W is by no means unique. There are infinitely many decompositions $W = \sum p_i' P[\psi']$ of W into nonorthogonal states ψ_i'. Indeed, by a linear transformation $\psi_i' = \sum b_{ij}\psi_j$ one can introduce a normalized, but nonorthogonal system $\{\psi_i'\}$ of states, i.e., states with $(\psi_i', \psi_i') = 1$. A state operator

$$W' = \sum_{i}^{\infty} p_i' P[\psi_i'], \qquad p_i' \geq 0, \qquad \mathrm{tr}\, W' = 1$$

is empirically equivalent to W if for any element $\chi \in \mathcal{H}$

$$\text{tr}\{WP[\chi]\} = \text{tr}\{W'P[\chi]\}, \qquad \forall \chi \in \mathcal{H}.$$

In this case, the expectation values of all observables with respect to the states W and W' are equal, and hence W and W' are empirically indistinguishable.

In this sense, there are infinitely many equivalent state operators $W^{(n)}$ that correspond to different nonorthogonal decompositions or mixtures of states $\Gamma^{(n)}(W) := \Gamma(p_i^{(n)}, \Psi_i^{(n)})$. (For a proof cf. [BeCa 81], pp. 9–11.) Hence, one could consider the state W as the formal representation of one of the situations given by a Gemenge $\Gamma^{(n)}(W)$. Since a given quantum system S cannot be simultaneously in different mixtures $\Gamma^{(n)}(W)$ of states, it follows that the state W is not sufficient to determine the Gemenge $\Gamma^{(n)}(W)$ that is actually realized. If W can be interpreted at all as the description of some Gemenge $\Gamma(W)$, new arguments must be added for a complete determination of the Gemenge $\Gamma(W)$ in question.

(ii) Under these conditions, one can argue in the following way. Even if the decomposition of a state W into pure states is generally not unique, there is one exception. The spectral decomposition of W into orthogonal states ψ_i is uniquely defined, so that one could assume that a mixed state W is the formal description of the 'spectral Gemenge' $\Gamma(p_i, \psi_i)$. The other decompositions would then be possibilities that are purely formal and without any physical relevance.

However, this way of reasoning cannot be correct. If the state W has a degenerate spectrum, the spectral decomposition reads

$$W = \sum p_i P(p_i), \qquad n_i = \text{tr}\, P(p_i)$$

where $P(p_i)$ projects onto the n_i-dimensional subspace M_i of states ψ_i^λ ($\lambda = 1, \ldots, n_i$) that correspond to the eigenvalue p_i of W. Even if one were to restrict consideration to the spectral decomposition into orthogonal components $P(p_i)$, in any subspace M_i one can introduce infinitely many orthogonal systems $\{\psi_i^{\lambda,\mu}\}$ ($\mu = 1, \ldots, \infty; \lambda = 1, \ldots, n_i$). Consequently, there are infinitely many decompositions

$$W^{(\mu)} = \sum_i p_i \sum_\lambda P[\psi_i^{\lambda,\mu}]$$

of W into orthogonal pure states $\psi_i^{\lambda,\mu}$. This means that even if one assumes that the spectral decomposition has some empirical preference, the decomposition of a mixed state W into orthogonal pure states is not unique and not determined by the state W itself. An interpretation of the mixed state W as a description of a mixture of orthogonal pure states is thus exposed to the objection that there is a large ambiguity of ensembles $\Gamma(p_i, \psi_i^{\lambda, \mu})$ with orthogonal pure states. However, this argument does not exclude the possibility that the system S which is described by the state W is actually in a pure state φ_i with probability p_i, but that the knowledge of W is not sufficient for the determination of the Gemenge $\Gamma(p_i, \varphi_i)$ which is described by W.

The arguments (i) and (ii) show that in the general case a mixed state W does not determine a mixture $\Gamma(p_i, \varphi_i)$ of states, even if one restricts the consideration to ensembles $\Gamma(p_i, \varphi_i)$ of orthogonal states. One could, however, try to assume that in spite of the above ambiguity the system S is actually in a well-defined Gemenge $\Gamma(p_i, \varphi_i)$, and that the observer does not know this Gemenge but only the mixed state W. We called this the 'ignorance interpretation' of the mixed state W. In the following subsection, we will investigate the question whether a given mixed state W admits an ignorance interpretation.

4.3(b) The nonobjectification theorems

According to the preceding discussion, the question whether a mixed state W admits an ignorance interpretation becomes relevant if the mixed state W was prepared by separation. Hence, for a discussion of this question let us consider a compound system $S = S_1 + S_2$ in a pure state $\Psi \in \mathcal{H}_S = \mathcal{H}_1 \otimes \mathcal{H}_2$ that is not of the product form $\varphi \otimes \Phi$ with $\varphi \in \mathcal{H}_1$ and $\Phi \in \mathcal{H}_2$. Let

$$\Psi = \sum c_i \, \varphi_i \otimes \Phi_i \tag{4.26}$$

be a biorthogonal decomposition of Ψ with orthonormal states $\varphi_i \in \mathcal{H}_1$, $\Phi_i \in \mathcal{H}_2$. It is well known that for any compound state $\Psi \in \mathcal{H}_1 \otimes \mathcal{H}_2$ there exists a biorthogonal decomposition with orthonormal systems $\{\varphi_i\}$ in $\mathcal{H}_1$ and $\{\Phi_i\}$ in $\mathcal{H}_2$. ([Neu 32], p. 231) The subsystems S_1 and S_2 are then in the reduced mixed states

$$W_1 = \sum_i |c_i|^2 \, P[\varphi_i] \tag{4.27a}$$

and

$$W_2 = \sum_i |c_i|^2 \, P[\Phi_i] \tag{4.27b}$$

respectively. We should repeat the sense in which it can be said that the subsystem S_1 is in the state W_1 and subsystem S_2 is in the state W_2. The compound system is always in the state $\Psi(S_1 + S_2)$. However, if one considers only those observables $\hat{A} = A_1 \otimes \mathbb{1}_2$ that refer to the degrees of freedom of the system S_1, say, then one can argue in the following way. The expectation values of all observables of this kind in the state $\Psi(S_1 + S_2)$ are equal to the expectation values of the S_1-observable A_1 in the mixed state $W_1(S_1)$. Hence, the subsystem S_1 can be treated *as if* it were prepared in the mixed state W_1. However, this way of speaking is limited to observables of the special form $A_1 \otimes \mathbb{1}_2$. The other observables of $S_1 + S_2$, which represent the correlations between S_1 and S_2, cannot be described in this way. This limitation will become important in the following discussion.

Let us consider now the situation mentioned above. A compound system $S = S_1 + S_2$ with subsystems S_1 and S_2 is in a pure state $\Psi(S_1 + S_2)$ and the subsystems S_1 and S_2 are in the mixed states W_1 and W_2, respectively. Using the basis systems $\{\varphi_i\}$ in $\mathscr{H}_1$ and $\{\Phi_i\}$ in $\mathscr{H}_2$, which appear in the biorthogonal decomposition of Ψ, the mixed state W_1, say, reads

$$W_1 = \sum_i |c_i|^2 \, P[\varphi_i].$$

If $A_1 = \sum a_i P[\varphi_i]$ is an observable in $\mathscr{H}_1$ with eigenstates φ_i and eigenvalues a_i, we can now formulate the following weak objectification assumption.

Weak ignorance interpretation of W_1. To the system S_1 in the mixed state W_1 one can assign hypothetically a value a_i of A_1 such that the value a_i pertains to the system with probability $p(W_1, a_i) = |c_i|^2$.

In this hypothesis, it is assumed that the quantum system S_1 actually possesses some value a_i of A_1, which, however, is not known to the observer, who knows only the probabilities $p(W_1, a_i)$.

If the observable A_1 can be weakly objectified with respect to S_1 in state W_1, then the observable $\hat{A} = A_1 \otimes \mathbb{1}_2 = \sum a_i P[\varphi_i] \otimes \mathbb{1}_2$ can be weakly objectified with respect to the compound system $S_1 + S_2$ in the state $\Psi(S_1 + S_2) = \sum_i c_i \varphi_i \otimes \Phi_i$. The weak objectification of $\hat{A}$ in the pure state Ψ can be treated in the same way as in section 4.2(c). A Lüders premeasurement of $\hat{A}$ on the system $S_1 + S_2$ in the state Ψ would lead to the mixed state

$$W_L(\Psi, \hat{A}) = \sum_i (P[\varphi_i] \otimes \mathbb{1}_2) P[\Psi](P[\varphi_i] \otimes \mathbb{1}_2)$$

$$= \sum_i |c_i|^2 P[\varphi_i \otimes \Phi_i]. \tag{4.28}$$

According to the arguments in section 4.2(c), the weak objectification of $\hat{A}$ in Ψ implies that for an arbitrary observable $\hat{B} = \sum \hat{B}_k P[\psi_k]$, $\psi_k \in \mathscr{H}_1 \otimes \mathscr{H}_2$, of the compound system $S_1 + S_2$ the probability of $\hat{B}_k$, $p(\Psi, \hat{B}_k) = \mathrm{tr}\{P[\Psi]P[\psi_k]\}$, must first fulfil in accordance with (4.20) the inequality

$$p(\Psi, \hat{B}_k) \leq \text{tr}\{W_L(\Psi, \hat{A}) P[\psi_k]\}.$$

However, since this inequality must hold for arbitrary test observables $\hat{B}$ and all values of k, one obtains also the inequality (as in (4.21))

$$p(\Psi, \hat{B}_k) \geq \text{tr}\{W_L(\Psi, \hat{A}) P[\psi_k]\}$$

and thus the equation

$$p(\Psi, \hat{B}_k) = \text{tr}\{W_L(\Psi, \hat{A}) P[\psi_k]\}, \qquad (4.29)$$

which is equivalent to the corresponding strong objectification condition in the sense of (4.3).

Since, in contrast to this result, quantum mechanics leads to the probability

$$p(\Psi, \hat{B}_k) = \text{tr}\{W_L(\Psi, \hat{A}) P[\psi_k]\} + p_{\text{int}}(\Psi; \hat{B}_k, \hat{A}) \qquad (4.30)$$

with the interference term

$$p_{\text{int}}(\Psi; \hat{B}_k, \hat{A}) = \text{tr}\Big\{\sum_{i \neq j} P[\varphi_i \otimes \Phi_i] P[\Psi] P[\varphi_j \otimes \Phi_j] P[\psi_k]\Big\}$$

it follows that eq. (4.29) is violated whenever $p_{\text{int}}(\Psi; \hat{B}_k, \hat{A}) \neq 0$. Consequently, we find that weak objectification of A_1 in the mixed state W_1 leads to a contradiction with the interference phenomena of the compound system, and hence the weak ignorance interpretation of the reduced mixed state must be rejected.

We summarize this result as the following nonobjectification (*NO*) theorem:

(*NO*) **theorem I.** *Let S be a quantum system with Hilbert space $\mathcal{H}$ and let $A = \sum a_i P[\varphi_i]$ be a discrete nondegenerate observable of S with a complete orthonormal system $\{\varphi_i\}$ in $\mathcal{H}$. If S is prepared in a mixed state $W = \sum p_i P[\varphi_i]$, $0 \leq p_i \leq 1$, by separation, then it is in general not possible to attribute a value a_i of A with probability p_i to the system S, i.e., the state W does not admit a weak ignorance interpretation.*

It is an obvious consequence of this (*NO*) theorem that the state $W = \sum p_i P[\varphi_i]$ also does not admit a strong ignorance interpretation. In order to make this result more explicit, let us assume that the subsystem S_1 with state W_1 in (4.27(a)) is actually in a state φ_i with probability $|c_i|^2$. The compound system $S_1 + S_2$ would then be in the product state $\varphi_i \otimes \Phi_i$ with probability $|c_i|^2$. Hence, $S_1 + S_2$ would be in the mixed state $W_L(\Psi, \hat{A})$, (4.28), and not in the pure state Ψ, (4.26). From (4.28) it follows that the probability for an arbitrary observable $\hat{B}$ is given by (4.29). Since this equation is in contradiction with the interference phenomena, the state W_1 does not admit a strong ignorance interpretation.

We summarize this result in the second nonobjectification (NO) theorem:

(NO) theorem II. *Let S be a quantum system with Hilbert space $\mathcal{H}$. If S is prepared in a mixed state W by separation, where $W = \sum_i p_i P[\varphi_i]$ is the spectral decomposition of W into orthogonal pure states φ_i, then it is in general not possible to assume that S is actually in one of the pure states φ_i with probability p_i, i.e., the state W does not admit a strong ignorance interpretation.*

Summarizing the content of the present subsection (4.3(b)), we find that a mixed state that is prepared by separation admits neither a strong nor a weak ignorance interpretation. Except in the case when a mixed state is prepared as a Gemenge, one cannot attribute to a proper quantum system in a mixed state W either an eigenstate φ_i of W or a value a_i of an observable A that has the same eigenstates φ_i as W. In this way, questions III and IV of Table 4.1 at the end of section 4.1 are answered in a negative sense.

4.3(c) Application to the measuring process

There are many interesting illustrations of the (NO) theorems I and II, e.g. the Einstein–Podolsky–Rosen experiment in the version of Bohm and Aharonov [BoAh 57], which is discussed in subsection 4.3(d). For the interpretation of quantum mechanics, the most important application of the (NO) theorems is to the quantum mechanical measuring process. Here we apply the formulation of the measuring process that was given in chapter 2.

The object system S and the measuring apparatus M are considered as proper quantum systems with initial states given by φ and Φ, respectively. In the first step of the measuring process, the preparation, the systems S and M are considered as a compound system $S + M$ in the state $\Psi(S + M) = \varphi(S) \otimes \Phi(M)$. In the second step, the premeasurement, the systems S and M interact, which is described here by a unitary operator $U(t)$ acting on the state $\Psi(S + M)$ within the time interval $0 \leq t \leq t'$. Hence, the compound state after the premeasurement is

$$\Psi'(S + M) = U(t')\Psi.$$

The state Ψ' and, hence, the interaction U are further specified by the calibration postulate (C_R), say, if we consider repeatable measurements. If

$A = \sum a_i P[\varphi^{a_i}]$ is the observable of S to be measured, then the calibration postulate reads

$$U(\varphi^{a_i} \otimes \Phi) = \varphi^{a_i} \otimes \Phi_i.$$

Here the states Φ_i are the orthonormal eigenstates of the pointer observable $Z = \sum Z_i P[\Phi_i]$. By means of the expansion $\varphi = \sum c_i \varphi^{a_i}$, with $c_i = (\varphi^{a_i}, \varphi)$, we can express in the general case the state of the compound system $S + M$ after the interaction in the form

$$\Psi' = U(\varphi \otimes \Phi) = \sum c_i \varphi^{a_i} \otimes \Phi_i.$$

The systems S and M can then be described by the reduced mixed states

$$W_S' = \sum |c_i|^2 P[\varphi^{a_i}], \qquad W_M' = \sum |c_i|^2 P[\Phi^{a_i}].$$

The third step of the measuring process, objectification and reading, assumes that these mixed states W_S' and W_M' admit an ignorance interpretation. The pointer would then possess a value Z_i with probability $p(\varphi, a_i) = |c_i|^2$. In order to discuss the question whether the mixed states W_S' and W_M' admit weak objectification of the observables A and Z respectively, we can apply (NO) theorems I and II of the preceding section. Indeed, since the mixed states W_S' and W_M' are prepared by separation, the premises mentioned in (NO) theorems I and II are fulfilled in both cases. An application of these theorems then leads to the following more specific result:

$(NO)_S$ **theorem** (Nonobjectification of the system observable). *It is in general not possible to attribute a value a_i of the observable A to the system S in the mixed state W_S', i.e., the state W_S' does not admit either a strong or a weak ignorance interpretation.*

This nonobjectification theorem $(NO)_S$ is of particular importance for the realistic interpretation I_R of quantum mechanics. The $(NO)_S$ theorem shows that by means of unitary repeatable premeasurements the system objectification postulate (SO) cannot be fulfilled in general. Since, however, the system objectification (SO) is one of the basic requirements of the realistic interpretation, it follows that quantum mechanics contradicts this part of the realistic interpretation I_R. The self-referential inconsistency between quantum mechanical object theory and its realistic interpretation, which is induced by the $(NO)_S$ theorem, is illustrated schematically in Fig. 4.4. Quantum mechanics implies the $(NO)_S$ theorem, which plainly contradicts the system objectification postulate (SO) assumed in the quantum theory of measurement.

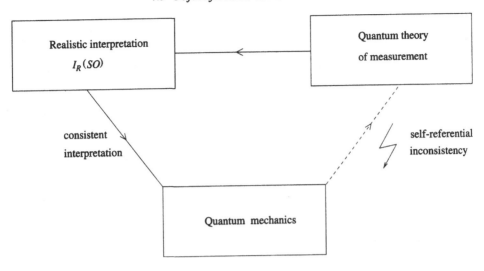

Fig. 4.4 Self-referential inconsistency between quantum mechanics and the system objectification postulate (SO) of the realistic interpretation I_R.

On the basis of this result, one could guess that the realistic interpretation of quantum mechanics, which makes use of repeatable measurements, is untenable, and one must confine oneself to the minimal interpretation I_M. However, this is not possible, since the (NO) theorems I and II can also be applied to the apparatus state W'_M, leading thus to the following result:

$(NO)_M$ **theorem.** Nonobjectification of the pointer observable. *It is in general not possible to attribute a value Z_i of the pointer observable Z to the apparatus M in the mixed state W'_M, i.e., the state W'_M does not admit either a strong or a weak ignorance interpretation.*

This nonobjectification theorem $(NO)_M$ shows that unitary premeasurements are even incompatible with the pointer objectification postulate (PO). Since, however, pointer objectification is an inevitable requirement of the minimal interpretation, one finds that quantum mechanics also contradicts the minimal interpretation I_M. The self-referential inconsistency between quantum mechanical object theory and its minimal interpretation, which is induced by the $(NO)_M$ theorem, is illustrated schematically in Fig. 4.5.

The contradiction between the quantum theory of measurement and the requirement of pointer objectification is one of the most serious problems in the foundations of quantum mechanics. We will come back to this topic, sometimes called the 'measurement problem' or the 'disaster of objectification' [Fraa 90] in a wider context in chapter 5.

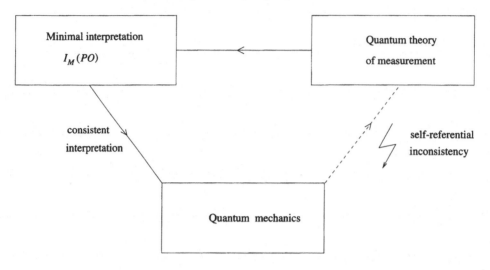

Fig. 4.5 Self-referential inconsistency between quantum mechanics and the pointer objectification postulate (PO) of the minimal interpretation I_M.

4.3(d) Illustration by spin experiments

The (NO) theorems I and II for mixed states can be illustrated and tested experimentally by means of two correlated spin-$\frac{1}{2}$ systems. The experiments discussed here were first proposed by Einstein, Podolsky, and Rosen [EPR 35] in a slightly different version and later reformulated by Bohm and Aharonov [BoAh 57] for two spin-$\frac{1}{2}$ systems. There are many experimental realizations of these Gedanken experiments by means of particles and photons. We mention here in particular the experiments by Aspect *et al.* [Asp 82], which made use of correlated polarized photons. It should be emphasized that the intentions of the papers mentioned were different from the aims of the present chapter. (We will come back to these problems in section 4.3.) The above-mentioned experiments, however, can also be used for a refutation of the ignorance interpretation of mixed states.

Let us consider two spin-$\frac{1}{2}$ systems S_1 and S_2 with two-dimensional Hilbert spaces $\mathscr{H}_1$ and $\mathscr{H}_2$ respectively and assume that the compound system $S = S_1 + S_2$ is in a pure state $\Psi(S) \in \mathscr{H}_1 \otimes \mathscr{H}_2$ that is given by

$$\Psi = \frac{1}{\sqrt{2}}(\varphi_{\mathbf{n}}^{(1)} \otimes \varphi_{-\mathbf{n}}^{(2)} - \varphi_{-\mathbf{n}}^{(1)} \otimes \varphi_{\mathbf{n}}^{(2)}) \qquad (4.31)$$

We denote the spin observables of the systems S_1 and S_2 by $\sigma_1(\mathbf{n})$ and $\sigma_2(\mathbf{n})$ where the spin direction is characterized by a unit vector $\mathbf{n}$ in the Poincaré sphere, as shown in Fig. 4.6. The eigenstates of $\sigma_1(\mathbf{n})$, say, are then given by $\varphi_{+\mathbf{n}}^{(1)}$ and $\varphi_{-\mathbf{n}}^{(1)}$ and correspond to eigenvalues $+1$ and -1 respectively. They

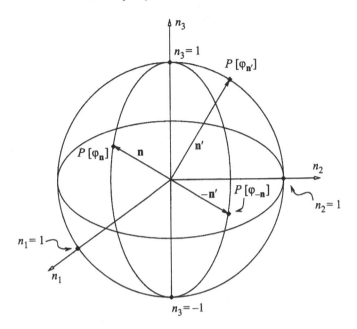

Fig. 4.6 Poincaré sphere for a spin-$\frac{1}{2}$ system. Pure states $\varphi_{\pm n}$ and projection operators $P[\varphi_{\pm n}]$ are given by points on the surface of the unit sphere. They can be represented by unit vectors $\mathbf{n} = (n_1, n_2, n_3)$ with $n_1^2 + n_2^2 + n_3^2 = 1$.

fulfil the eigenvalue equations

$$\sigma_1(\mathbf{n})\varphi_{\pm n}^{(1)} = \pm\varphi_{\pm n}^{(1)}.$$

The states of the subsystems S_1 and S_2 are given here by the reduced mixed states

$$W_1(S_1) = \mathrm{tr}_2 P[\Psi] = \tfrac{1}{2}P[\varphi_{+n}^{(1)}] + \tfrac{1}{2}P[\varphi_{-n}^{(1)}] = \tfrac{1}{2}\mathbb{1}_1,$$

$$W_2(S_2) = \mathrm{tr}_1 P[\Psi] = \tfrac{1}{2}P[\varphi_{+n}^{(2)}] + \tfrac{1}{2}P[\varphi_{-n}^{(2)}] = \tfrac{1}{2}\mathbb{1}_2.$$

We can now apply the weak and strong ignorance interpretation hypotheses to the system S_1, say, in the state W_1. The weak objectification hypothesis for the spin observable $\sigma_1(\mathbf{n})$ then reads:

To the system S_1 in the state W_1 one can assign a value $s_i \in \{\pm 1\}$ of $\sigma_1(\mathbf{n})$ such that the value s_i pertains to the system with probability $\frac{1}{2}$.

This weak objectification of $\sigma_1(\mathbf{n})$ in S_1 has the consequence that the observable $\hat{A}(\mathbf{n}) = \sigma_1(\mathbf{n}) \otimes \mathbb{1}_2$ is weakly objectified with respect to the compound system $S_1 + S_2$ in the state $\Psi \in \mathscr{H}_1 \otimes \mathscr{H}_2$. An (ideal) Lüders premeasurement of $\hat{A}(\mathbf{n})$ on the system S in the state Ψ would then lead to the mixed state

$$W_L(\Psi, \hat{A}(\mathbf{n})) = \tfrac{1}{2}P_{+-}(\mathbf{n}, \mathbf{n}) + \tfrac{1}{2}P_{-+}(\mathbf{n}, \mathbf{n}) \qquad (4.32)$$

with the notation

$$P_{ik}(\mathbf{n}', \mathbf{n}'') := P[\varphi_{i\mathbf{n}'}^{(1)} \otimes \varphi_{k\mathbf{n}''}^{(2)}]. \qquad (4.33)$$

According to the arguments in subsection 4.3(b), the weak objectification of $\hat{A}(\mathbf{n})$ with respect to $S_1 + S_2$ in the state Ψ implies that for an arbitrary test observable $\hat{B} = \sum \hat{B}_i P[\Psi_i]$ of the compound system the probabilities

$$p(\Psi, \hat{B}_k) = \mathrm{tr}\{P[\Psi]P[\Psi_k]\}$$

must fulfil – in accordance with (4.29) – the weak objectification condition

$$p(\Psi, \hat{B}_k) = \mathrm{tr}\{P[\Psi]P[\Psi_k]\} = \mathrm{tr}\{W_L(\Psi, \hat{A}(\mathbf{n}))P[\Psi_k]\}, \qquad (4.34)$$

which is equivalent to the corresponding strong objectification condition.

Here we make use of the special test observable $\hat{B}$,

$$\hat{B}(\mathbf{n}', \mathbf{n}'') := \sigma_1(\mathbf{n}') \otimes \sigma_2(\mathbf{n}''), \qquad (4.35)$$

with spectral decomposition

$$\hat{B}(\mathbf{n}', \mathbf{n}'') = \sum_{i,k \in \{+,-\}} \hat{B}_{ik} P_{ik}(\mathbf{n}', \mathbf{n}'') \qquad (4.36)$$

and eigenvalues $\hat{B}_{++} = \hat{B}_{--} = 1$, $\hat{B}_{+-} = \hat{B}_{-+} = -1$. If one assumes that $\hat{A}(\mathbf{n})$ is weakly objectified with respect to the compound system S in the state Ψ, then for the probabilities of the eigenvalues $\hat{B}_{ik}$ ($i, k \in \{+, -\}$) of $\hat{B}$ one obtains according to (4.34) and (4.36)

$$p(\Psi, \hat{B}_{ik}) = \mathrm{tr}\{W_L(\Psi, \hat{A}(\mathbf{n}))P_{ik}(\mathbf{n}', \mathbf{n}'')\}. \qquad (4.37)$$

For the special choice of Ψ, $\hat{A}$, $W_L(\Psi, \hat{A})$, and $\hat{B}$ given by (4.31)–(4.33), (4.35), (4.36) it follows that

$$\mathrm{tr}\{P[\Psi]P_{ik}(\mathbf{n}', \mathbf{n}'')\} = \tfrac{1}{4}(1 - \hat{B}_{ik}(\mathbf{n}' \cdot \mathbf{n}'')),$$

$$\mathrm{tr}\{W_L(\Psi, \hat{A}(\mathbf{n}))P_{ik}(\mathbf{n}', \mathbf{n}'')\} = \tfrac{1}{4}(1 - \hat{B}_{ik}(\mathbf{n} \cdot \mathbf{n}')(\mathbf{n} \cdot \mathbf{n}'')).$$

Hence the condition (4.37) assumes for all values $\hat{B}_{ik}$ the special form

$$\mathbf{n}' \cdot \mathbf{n}'' - (\mathbf{n} \cdot \mathbf{n}')(\mathbf{n} \cdot \mathbf{n}'') \equiv (\mathbf{n} \times \mathbf{n}')(\mathbf{n} \times \mathbf{n}'') = 0. \qquad (4.38)$$

This equation is the condition under which the observable $\hat{A}(\mathbf{n})$ can be weakly objectified in quantum mechanics with respect to the state Ψ if the observables $\hat{B}(\mathbf{n}', \mathbf{n}'')$ are used as test observables.

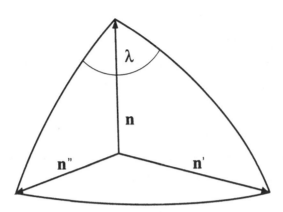

Fig. 4.7 Unit vectors **n**, **n**′, and **n**″ representing the objectified observable $\hat{A}(\mathbf{n})$ and the test observable $\hat{B}(\mathbf{n}', \mathbf{n}'')$.

The weak objectification condition (4.38) is fulfilled if

1. $\mathbf{n} \times \mathbf{n}' = 0$, i.e., **n** and **n**′ are parallel, or
2. $\mathbf{n} \times \mathbf{n}'' = 0$, i.e., **n** and **n**″ are parallel, or
3. the planes spanned by $(\mathbf{n}, \mathbf{n}')$ and $(\mathbf{n}, \mathbf{n}'')$ are orthogonal, i.e., $\cos \lambda = 0$ (Fig. 4.7).

Cases 1 and 2 mean that the test observable $\hat{B}$ partially coincides with the objectified observable $\hat{A}$. If arbitrary test observables are admitted, then the weak objectification condition can be violated. This means that weak objectification of the observable $A(\mathbf{n})$ with respect to the system S_1 in the mixed state W_1 is not compatible with quantum mechanics.

An experimental confirmation of this result has to show that for a two-spin-$\frac{1}{2}$ system in the singlet state (4.31) the statistics of measurement outcomes for a given test observable reproduces the quantum mechanical probabilities – and not the probabilities (4.37) which follow from the weak objectification hypothesis.

This means in particular, that quantum mechanical interference experiments are needed to disprove the ignorance interpretation hypothesis (4.37). Experiments with correlated polarized photons, which are discussed in the following section, can also be used for a falsification of the weak objectification hypothesis.

4.4 Probability attribution

4.4(a) From Kant to Boole

In his *Treatise of Human Nature* (1739) David Hume emphasized that we never observe objects but only qualities and that it is nothing but imagination if we regard the observed qualities as properties of an object. Hence any scientific cognition begins with the observation of qualities and it seems to be merely a question of interpretation whether in addition to the observed phenomena a fictitious object, 'an unknown something', is used for the description of the experimental results. Obviously, there is no reason to expect that general laws like conservation of substance or causality hold in nature.

The same problem was treated by Kant in the *Critique of Pure Reason* (1787). However, in contrast to Hume, Kant emphasized that 'objects of experience' are not arbitrary imaginings but entities that are constituted from the observational data by means of some well-defined conceptual prescriptions, the categories of substance and causality. Hence, the interpretation of the empirical data as properties of an object can only be justified if the object as carrier of properties is constituted by these categories. Kant formulated necessary conditions that must be fulfilled by the observational data if the measurement results are to be considered as properties of an 'object of experience'.[†]

This way of reasoning can easily be illustrated within the framework of classical mechanics. If we possess some objective cognition that relates to the material external reality and not to the observing subject, then the observations in space and time must be connected by mechanical laws, which are specifications and realizations of the general laws of causality and of the conservation of substance. For example, in the planetary system there is a large number of small planets that can be observed only occasionally. Observations that can be obtained within a period of several months, say, consist of many isolated light points without any obvious connection. However, if these observed data refer to a well-defined astronomical object, then the light points must be points on a space-time trajectory that is determined by Newton's equation of motion. This mechanical law is strictly causal and

---

† It should be emphasized that Kant did not claim that an interpretation of this kind is possible for any set of observations. However, 'if each representation were completely foreign to every other, standing apart in isolation, no such thing as knowledge would ever arise. For knowledge is [essentially] a whole in which representations stand compared and connected'. In other words, without causal correlation the observed qualities cannot be considered as properties of an object of experience ([Kant 1787], A 97). This case of 'empty knowledge' will become interesting also for the problem of probability attribution.

it conserves substance, i.e., it preserves the mass point as the carrier of the mechanical predicates.

The physical sciences such as classical mechanics, which are governed by strict causality laws, were extended in the nineteenth century by the incorporation of statistical theories. In these new fields, the primarily given entities are no longer events E_λ in space and time whose causal connection can easily be recognized but probabilities $p(E_\lambda)$ for the occurrence of events E_λ or, more correctly, the relative frequencies of the events E_λ. Under these conditions the original Kantian way of reasoning is no longer applicable. If causal correlations between several events are not discernible, the objectivity of our cognition cannot be guaranteed in the Kantian way. Instead, one has to begin with the only perceptible empirical structure, the relative frequencies or probabilities $p(E_\lambda)$ of various events E_λ, and ask under which conditions probabilities of this kind refer to a certain exterior object.

This problem was first studied by George Boole, best known as one of the founders of modern logic. In his book *The Laws of Thought* [Boole 1854] and in a subsequent paper *On the Theory of Probabilities* [Boole 1862], Boole investigated what he called the 'condition of possible experience'. The problem treated by Boole can be sketched as follows. Assume that the observer is given a set of numbers $p_1, p_2, \ldots, p_n$ with $0 \le p_i \le 1$, which represent the relative frequencies of n merely *logically connected* events $E_1, E_2, \ldots, E_n$. One can then ask for the necessary and sufficient conditions that must be fulfilled if the numbers p_i are to be considered as probabilities for properties (given by E_i) of some object of experience. This problem will be discussed in detail in the following section (cf. also [Pit 94]).

4.4(b) Classicality condition

In order to realize Boole's project, let us consider a sequence of n rational numbers

$$p_1, p_2, \ldots, p_n,$$

which correspond to the observed relative frequencies of the logically connected events $E_1, E_2, \ldots, E_n$. 'Logically connected' means in this context that the events E_i are not logically independent but connected by logical operations and relations. The question that will be investigated here reads: what are the necessary and sufficient conditions that permit the interpretation of these numbers p_i as 'probabilities' that can be attributed to some physical system? In other words, what are the conditions for the existence of a classical probability space such that there are n logically connected events

$E_1, E_2, \ldots, E_n$ whose probabilities $p(E_i)$ are given by the observed numbers p_i?

If the events E_i are logically independent, i.e., if they are not logically connected in any way, then the only conditions that must be fulfilled by the probabilities p_i are

$$0 \le p_i \le 1. \tag{4.39}$$

Clearly, these conditions are always fulfilled by the observed relative frequencies. A situation in which the observed data – the events E_i – do not show any obvious connection in either a causal or a logical sense can be compared with the 'empty knowledge' case mentioned by Kant. Indeed, if each event were completely isolated from every other, 'no such thing as knowledge would ever arise'. In this case, the numbers p_i $(0 \le p_i \le 1)$ could not be attributed to an object system as the probabilities of its various properties.

If, however, the events E_i are logically interconnected, then one can investigate the question of whether the observed relative frequencies $p_i = f(E_i)$ of the events E_i can be attributed to an object system as probabilities for the properties given by E_i. For a more rigorous treatment of this problem, one has to specify first the logic of the events and second the probability measure in question. Here we will give a brief account of the basic concepts needed for the following discussion. For more details and formal proofs we refer to the literature, in particular, to the work of Beltrametti and Maczynski [BeMa 91] and of Pitowsky [Pit 89].

An *event system L* is a triple $L = \langle L_0, \le, \neg \rangle$ where L_0 is a set of (elementary) events that is equipped with a partial ordering relation $\le$ and a 1-place operation $\neg$, the complementation, such that L is an orthocomplemented orthomodular partially ordered set which can be extended to a lattice by means of the 2-place operations $\wedge$ and $\vee$. In particular a system L is called a *classical event system* if it is a complemented distributive lattice with respect to the relation $\le$ and the complementation $\neg$. Clearly, the logical meaning of the relation $\le$ is the implication, the operation $\neg$ is the negation, and the operations $\wedge$ (and) and $\vee$ (or) can be defined as infimum and as supremum, respectively.

A *probability measure p* on the event system L is a map

$$p : L \to [0, 1]$$

of the set L of events onto the interval $[0, 1]$ which satisfies the

Kolmogorov axioms

$$(i) \quad p(\mathbf{0}) = 0, \qquad p(\mathbf{I}) = 1$$

$$(ii) \quad p(\neg a) = 1 - p(a) \quad \text{for all } a \in L \tag{4.40}$$

$$(iii) \quad p(a_1 \vee a_2 \vee \cdots \vee a_n) = p(a_1) + p(a_2) + \cdots + p(a_n)$$

$$\text{whenever } a_i \leq \neg a_j \text{ for } i \neq j.$$

By an *event probability space* W we understand a pair $W = \langle L, p \rangle$, where L is an event system and p a probability measure on L. By means of the concepts L, p, and W we can now formulate the main problem.

A correlation sequence $K = \{p_1, p_2, \ldots, p_n; p_{ij}, \ldots\}$ is a set of elementary probabilities p_i and joint probabilities p_{ij}, where not all pairs (i, j) need to appear. A sequence K is called *consistently representable* if there is an event probability space $W = \langle L, p \rangle$ and a sequence of events $(a_1, a_2, \ldots, a_n)$ in L such that

$$p_i = p(a_i), \qquad p_{ij} = p(a_i \wedge a_j)$$

whenever the pair (i, j) appears in K.

In particular, K is called classically representable if L is a classical event system. This means that the elements of K – the observed relative frequencies – may be considered as classical (Kolmogorovian) probabilities that refer to the elements a_i of a classical event system L_B. The property of a correlation sequence K of being consistently representable in a classical event space is called 'classicality'. The (necessary and sufficient) conditions under which a probability sequence K possesses the property of classicality are called 'classicality conditions'. The classicality conditions show that a given sequence of probabilities provides some objective knowledge about a physical system. Indeed, if we are given (by experiment) a sequence of probabilities $p_i = p(E_i)$ of events E_i, then the necessary and sufficient conditions for these events to indicate classical properties $P_i(S)$ of a physical system S are the classicality conditions.

4.4(c) Bell inequalities

In order to illustrate the concept of classicality, we consider two examples, probability sequences with two and three properties, respectively. For properties we write a, b, c, ... and for their probabilities $p(a)$, $p(b)$, $p(c)$, ..., respectively. For the joint probability (of the property $a \wedge b$) we write $p(a, b)$, etc. The simplest example of a probability sequence is then given by a sequence with two properties a and b:

$$K_2 = \{p(a), p(b), p(a, b)\}. \tag{4.41}$$

Classicality theorem I. *The correlation sequence K_2 is consistently representable in a classical event space if and only if the 'classicality conditions'*

$$0 \le p(a, b) \le p(a) \le 1$$
$$0 \le p(a, b) \le p(b) \le 1 \tag{4.42}$$
$$p(a) + p(b) - p(a, b) \le 1$$

hold.

In this theorem we have considered only positive properties. If one introduces the negations $\neg a$ and $\neg b$ of a and b respectively, with $p(\neg a) = 1 - p(a)$, $p(\neg b) = 1 - p(b)$, $p(\neg a, \neg b) = 1 - p(a) - p(b) + p(a, b)$ then one finds that the inequalities (4.42) are necessary and sufficient for the following relations:

$$0 \le p(\neg a, \neg b) \le p(\neg a) \le 1$$
$$0 \le p(\neg a, \neg b) \le p(\neg b) \le 1 \tag{4.43}$$
$$p(\neg a) + p(\neg b) - p(\neg a, \neg b) \le 1.$$

In the following both systems of inequalities will be used.

For the proof of classicality theorem I, one needs the probability axioms (4.40) and some properties of the Boolean lattice $L_B^{(2)}$ that are generated by the two elementary properties a and b. The two parts of the proof (that conditions (4.42) are necessary and sufficent) are shown here (in Appendix 6) explicitly, since they illustrate the concepts of 'classicality' and 'probability-attribution'.

The meaning of the classicality conditions (4.42) and (4.43) can be further illustrated by the following example, which is a slight generalization of the experiment discussed in subsection 4.2(b) [BLM 92]. Let us consider a quantum system S with a two-dimensional Hilbert space, the pure states of which are represented by the surface of a unit (Poincaré) sphere. In accordance with the terminology of the spin-$\frac{1}{2}$ system in Fig. 4.6, unit vectors are denoted here by $\mathbf{a}$, $\mathbf{b}$, $\mathbf{c}$, ... and the corresponding pure states and projection operators by $\varphi_{\mathbf{a}}$, ... and $P[\varphi_{\mathbf{a}}]$, ..., respectively (Fig. 4.6). Observables A, B, ... are then represented by their spectral projections $P[\varphi_{\pm \mathbf{a}}]$, $P[\varphi_{\pm \mathbf{b}}]$, ... If the quantum system is in a pure state $\varphi_{\mathbf{c}}$, the probabilities of $P[\varphi_{\pm \mathbf{a}}]$ and $P[\varphi_{\pm \mathbf{b}}]$ are given by

$$p(\varphi_{\mathbf{c}}; \pm \mathbf{a}) = \tfrac{1}{2}(1 \pm \mathbf{a} \cdot \mathbf{c}), \quad p(\varphi_{\mathbf{c}}; \pm \mathbf{b}) = \tfrac{1}{2}(1 \pm \mathbf{b} \cdot \mathbf{c}). \tag{4.44}$$

Remark. The photon split beam experiment of subsection 4.2(b) corresponds to the special case $\mathbf{a} \cdot \mathbf{b} = \mathbf{a} \cdot \mathbf{c} = 0$ and $\mathbf{b} \cdot \mathbf{c} = \delta$. Hence, we get $p(\varphi_c; \mathbf{a}) = \frac{1}{2}$ and $p(\varphi_c; \mathbf{b}) = \cos^2(\delta/2)$ in accordance with (4.7) and (4.8).

We compare the quantum mechanical probabilities (4.44) with the classicality condition (4.42) for two observables A and B and a system in the state φ_c. The classicality conditions then read

$$0 \leq p(\varphi_c; \mathbf{a}, \mathbf{b}) \leq p(\varphi_c; \mathbf{a}) \leq 1$$
$$0 \leq p(\varphi_c; \mathbf{a}, \mathbf{b}) \leq p(\varphi_c; \mathbf{b}) \leq 1$$
$$p(\varphi_c; \mathbf{a}) + p(\varphi_c; \mathbf{b}) - p(\varphi_c; \mathbf{a}, \mathbf{b}) \leq 1.$$

In the present case, all joint probabilities are zero, and hence the first two lines are trivially fulfilled. For the third line, we get

$$\tfrac{1}{2}(1 \pm \mathbf{a} \cdot \mathbf{c}) + \tfrac{1}{2}(1 \pm \mathbf{b} \cdot \mathbf{c}) \leq 1$$

and thus $\mathbf{a} \cdot \mathbf{c} + \mathbf{b} \cdot \mathbf{c} \leq 0$. Since the classicality conditions hold also for the negations $(-\mathbf{a}, -\mathbf{b})$, (4.43) yields $-\mathbf{a} \cdot \mathbf{c} - \mathbf{b} \cdot \mathbf{c} \leq 1$ and hence we have

$$\mathbf{a} \cdot \mathbf{c} + \mathbf{b} \cdot \mathbf{c} \equiv (\mathbf{a} + \mathbf{b}) \cdot \mathbf{c} = 0. \tag{4.45}$$

This equation is satisfied for states φ_c with $\mathbf{c} \perp (\mathbf{a} + \mathbf{b})$. In these cases the classicality conditions (4.42), (4.43) are in accordance with quantum mechanics. In all other cases, the classicality conditions are violated in quantum theory.

This result shows that in quantum mechanics the probability sequence K_2 cannot be represented in a classical event space, except when $\mathbf{c} \perp (\mathbf{a} + \mathbf{b})$, which means that probability attribution is in general not possible.

It is interesting to compare this result, i.e., (4.45), with the *weak objectification* requirement discussed in subsection 4.2(c). If we assume that A is weakly objectified in the state φ_c then, according to (4.20), for the probabilities of the values of the observable B, i.e., of the $P[\varphi_{\pm b}]$, we obtain the equations

$$p(\varphi_c, \pm \mathbf{b}) = \mathrm{tr}\{W_L(\varphi_c, A) P[\varphi_{\pm b}]\} \tag{4.46}$$

with

$$W_L(\varphi_c, A) = p(\varphi_c, \mathbf{a}) P[\varphi_{\mathbf{a}}] + p(\varphi_c, -\mathbf{a}) P[\varphi_{-\mathbf{a}}]$$
$$= \tfrac{1}{2}(1 + \mathbf{a} \cdot \mathbf{c}) P[\varphi_{\mathbf{a}}] + \tfrac{1}{2}(1 - \mathbf{a} \cdot \mathbf{c}) P[\varphi_{-\mathbf{a}}].$$

Equation (4.46) thus reads

$$\tfrac{1}{2}(1 \pm \mathbf{b} \cdot \mathbf{c}) = \tfrac{1}{2}(1 \pm (\mathbf{a} \cdot \mathbf{b})(\mathbf{a} \cdot \mathbf{c}))$$

and hence the weak objectification of A can be expressed by the relation

$$\mathbf{b} \cdot \mathbf{c} - (\mathbf{a} \cdot \mathbf{b})(\mathbf{a} \cdot \mathbf{c}) \equiv (\mathbf{a} \times \mathbf{b}) \cdot (\mathbf{a} \times \mathbf{c}) = 0. \qquad (4.47)$$

This equation is fulfilled if one of the factors is zero or otherwise if the planes spanned by $(\mathbf{a}, \mathbf{b})$ and $(\mathbf{a}, \mathbf{c})$ are orthogonal, i.e., $\cos \lambda = 0$ in Fig 4.7. The case $\mathbf{a} \times \mathbf{c} = 0$ means that φ_c is an eigenstate of A, whereas $\mathbf{a} \times \mathbf{b} = 0$ occurs if the test observable B coincides with the objectified observable A. If the state φ_c is not an eigenstate of A, then the weak objectification condition (4.47) is violated in quantum mechanics for all test observables B such that $B \neq A$ and $\cos \lambda \neq 0$.

The strength of the weak objectification condition (4.47), or of the classicality condition (4.45), corresponds to the size of the domains in the Poincaré sphere in which these conditions are violated. In these regions the conditions of weak objectification or of classicality are not in accordance with quantum mechanics. In (4.47) the vector $\mathbf{c}$ represents the preparation, $\mathbf{a}$ the objectified observable, and $\mathbf{b}$ the test observable. If, on the one hand, one allows for arbitrary test observables $\mathbf{b}$, then (4.47) is violated almost everywhere in the Poincaré sphere except for the trivial case $\mathbf{a} = \pm \mathbf{c}$, when φ_c is an eigenstate of A. On the other hand, in (4.45) the vectors $\mathbf{a}$ and $\mathbf{b}$ correspond to observables whose probabilities are attributed to the system in the sense of classicality. This equation is violated only for states φ_c for which $\mathbf{c} \not\perp (\mathbf{a} + \mathbf{c})$. Hence the requirement of probability attribution is weaker than the weak objectification condition. This special result confirms the general remarks made in subsection 4.1(a).

The most important case deals with *three* independent properties a, b, c for which the Boolean lattice $L_B^{(3)}$ is generated by the elements $\{a, b, c\}$ and the corresponding probabilities $p(a)$, $p(b)$, $p(a, b)$, etc. In this case, we have the following result:

Classicality theorem II. *Let* $K_3 = \{p(a), p(b), p(c); p(a, b), p(a, c), p(b, c)\}$ *be a correlation sequence of probabilities of events* a, b, c. *The correlation sequence* K_3 *is consistently representable in a classical event space if and only if the 'classicality conditions' hold (here we denote the three properties by* $(a_1, a_2, a_3) = (a, b, c)$):

$$0 \leq p(a_i, a_k) \leq p(a_i) \leq 1$$
$$0 \leq p(a_i, a_k) \leq p(a_k) \leq 1 \qquad (4.48a)$$
$$p(a_i) + p(a_k) - p(a_i, a_k) \leq 1$$

with $1 \leq i < k \leq 3$ *and in addition*

$$p(a_1) - p(a_1, a_2) - p(a_1, a_3) + p(a_2, a_3) \geq 0$$
$$p(a_2) - p(a_1, a_2) - p(a_2, a_3) + p(a_1, a_3) \geq 0 \qquad (4.48b)$$
$$p(a_3) - p(a_1, a_3) - p(a_2, a_3) + p(a_1, a_2) \geq 0$$
$$p(a_1) + p(a_2) + p(a_3) - p(a_1, a_2) - p(a_1, a_3) - p(a_2, a_3) \leq 1.$$

The second group of inequalities (4.48b) is known as the Bell inequalities and was first discovered by J. S. Bell in an investigation of hidden parameters in quantum mechanics [Bell 64]. In the present form, the inequalities (4.48b) appear first in the work of Pitowsky [Pit 89] and of Beltrametti and Maczynski [BeMa 91].

The inequalities (4.48b) are formulated in terms of positive properties. A formulation that is more familiar from the literature can be obtained by eliminating the marginal probabilities $p(a_i)$. Renaming (a_1, a_2, a_3) as (a, b, c) and using marginal relations like

$$p(a) = p(a, b) + p(a, \neg b)$$
$$p(a) = p(a, c) + p(a, \neg c)$$

we get

$$p(b, \neg c) \leq p(\neg a, b) + p(a, \neg c)$$
$$p(b, c) \leq p(\neg a, b) + p(a, c) \qquad (4.49)$$
$$p(\neg a, b) \leq p(\neg a, c) + p(b, \neg c)$$
$$p(\neg a, c) \leq p(\neg a, \neg b) + p(b, c).$$

The system (4.49) of Bell inequalities will be used in the following discussion. For the proof of the classicality theorem II, which is more complicated than the proof of classicality theorem I, we refer to theorem 2.3 in [Pit 89] and to theorem II in [BeMa 91]. Furthermore, it should be mentioned that the Bell inequalities (4.49) are also necessary and sufficient for the existence of a 3-joint probability distribution [Fine 82], [Bus 85].

4.4(d) Illustration by experiment

In order to illustrate the meaning of the Bell inequalities (4.49), we consider again the two spin-$\frac{1}{2}$ systems that were discussed in subsection 4.3(d). In this experiment, two correlated spin-$\frac{1}{2}$ systems S_1 and S_2 are prepared in such a way that the compound system $S = S_1 + S_2$ is in a pure singlet state Ψ given by

$$\Psi = \tfrac{1}{\sqrt{2}}(\varphi_{\mathbf{n}}^{(1)} \otimes \varphi_{-\mathbf{n}}^{(2)} - \varphi_{-\mathbf{n}}^{(1)} \otimes \varphi_{\mathbf{n}}^{(2)}) \tag{4.50}$$

in accordance with (4.31). The subsystems S_1 and S_2 are then in the reduced mixed states

$$W_1(S_1) = \tfrac{1}{2}\mathbb{1}_1, \qquad W_2(S_2) = \tfrac{1}{2}\mathbb{1}_2. \tag{4.51}$$

Let us consider the three different properties of system S_1 that are given by the projection operators

$$P[\varphi_{\mathbf{n}}^{(1)}], \quad P[\varphi_{\mathbf{n'}}^{(1)}], \quad P[\varphi_{\mathbf{n''}}^{(1)}]$$

corresponding to spin observables with directions $\mathbf{n}$, $\mathbf{n'}$, $\mathbf{n''}$, respectively (for illustration, see also Figs. 4.6 and 4.7).

The classicality conditions for these three properties are then given by the Bell inequalities (4.49). If we denote the probability of the property $P[\varphi_{\mathbf{n}}^{(1)}]$, say, by $p_1(\mathbf{n})$ and the joint probability corresponding to pairs $(\mathbf{n}, \mathbf{n'})$ by $p_1(\mathbf{n}, \mathbf{n'})$, etc., then the Bell inequalities (4.49) read

$$
\begin{aligned}
p_1(\mathbf{n''}, -\mathbf{n}) &\le p_1(-\mathbf{n'}, \mathbf{n''}) + p_1(\mathbf{n'}, -\mathbf{n}) \\
p_1(\mathbf{n''}, \mathbf{n}) &\le p_1(-\mathbf{n'}, \mathbf{n''}) + p_1(\mathbf{n'}, \mathbf{n}) \\
p_1(-\mathbf{n'}, \mathbf{n''}) &\le p_1(-\mathbf{n'}, \mathbf{n}) + p_1(\mathbf{n''}, -\mathbf{n}) \\
p_1(-\mathbf{n'}, \mathbf{n}) &\le p_1(-\mathbf{n'}, -\mathbf{n''}) + p_1(\mathbf{n''}, -\mathbf{n})
\end{aligned}
\tag{4.52}
$$

if one identifies (a, b, c) from (4.49) with $(\mathbf{n}, \mathbf{n'}, \mathbf{n''})$ [BLM 92].

Within the context of the Einstein–Podolsky–Rosen (EPR) situation, the joint probabilities $p_1(\mathbf{n}, \mathbf{n''})$, etc., that refer to the system S_1 can be determined experimentally, although the respective observables $P[\varphi_{\mathbf{n}}^{(1)}]$ and $P[\varphi_{\mathbf{n''}}^{(1)}]$ are not commensurable for $\mathbf{n} \ne \mathbf{n''}$. The two systems S_1 and S_2 are subsystems of the compound system $S = S_1 + S_2$ prepared in the pure singlet state Ψ given by (4.50). Hence, the systems S_1 and S_2 are in the reduced mixed states W_1 and W_2 as given in (4.51). The probability $p_1(\mathbf{n}, \mathbf{n'})$ that S_1 possesses jointly the properties which correspond to $\mathbf{n}$ and $\mathbf{n'}$ is then equal to the probability $p_{12}(\mathbf{n}, -\mathbf{n'})$ that the $\mathbf{n}$-property pertains to S_1 and the $(-\mathbf{n'})$-property to S_2. On account of the rotational symmetry of the compound state $\Psi(S_1 + S_2)$, one thus obtains, e.g.,

$$p_1(\mathbf{n}, -\mathbf{n'}) = p_{12}(\mathbf{n}, \mathbf{n'}) = p_{12}(-\mathbf{n}, -\mathbf{n'}). \tag{4.53}$$

If in an EPR experiment the two subsystems are sufficiently separated – as, e.g., in the experiment performed by Aspect *et al.* [Asp 82] – one can simultanously measure the $\mathbf{n}$-property on S_1 and the $\mathbf{n'}$-property on S_2. In this way, it is possible to determine experimentally $p_{12}(\mathbf{n}, \mathbf{n}')$ and thus also $p_1(\mathbf{n}, -\mathbf{n}')$.

Using the explicit form (4.50) of the compound state $\Psi(S_1 + S_2)$, one obtains the quantum mechanical value of $p_{12}(\mathbf{n}, \mathbf{n}')$:

$$p_{12}(\mathbf{n}, \mathbf{n}') = p(\Psi; \mathbf{n}, \mathbf{n}') = \tfrac{1}{4}(1 - \mathbf{n} \cdot \mathbf{n}'). \qquad (4.54)$$

If one replaces in the four Bell inequalities (4.52) the S_1-probabilities by their quantum mechanical values, e.g., $p_1(\mathbf{n}, \mathbf{n}')$ by $p(\Psi; \mathbf{n}, -\mathbf{n}')$, as given by (4.53) and (4.54), one obtains after some rearrangements the inequalities

$$
\begin{aligned}
\mathbf{n} \cdot \mathbf{n}' + \mathbf{n}' \cdot \mathbf{n}'' &\leq 1 + \mathbf{n} \cdot \mathbf{n}'' \\
\mathbf{n}' \cdot \mathbf{n}'' - \mathbf{n} \cdot \mathbf{n}' &\leq 1 - \mathbf{n} \cdot \mathbf{n}'' \\
\mathbf{n} \cdot \mathbf{n}' - \mathbf{n}' \cdot \mathbf{n}'' &\leq 1 - \mathbf{n} \cdot \mathbf{n}'' \\
-(\mathbf{n} \cdot \mathbf{n}' + \mathbf{n}' \cdot \mathbf{n}'') &\leq 1 + \mathbf{n} \cdot \mathbf{n}''
\end{aligned}
\qquad (4.55a)
$$

which can also be written in the more compact form

$$
\begin{aligned}
|\mathbf{n} \cdot (\mathbf{n}' - \mathbf{n}'')| &\leq \mathbf{n}' \cdot (\mathbf{n}' - \mathbf{n}'') \\
|\mathbf{n} \cdot (\mathbf{n}' + \mathbf{n}'')| &\leq \mathbf{n}' \cdot (\mathbf{n}' + \mathbf{n}'').
\end{aligned}
\qquad (4.55b)
$$

Triples of vectors $(\mathbf{n}, \mathbf{n}', \mathbf{n}'')$ which fulfil the inequalities (4.55) satisfy the Bell inequalities in accordance with quantum mechanics. Moreover, since the full set of Bell inequalities (4.52) is necessary and sufficient for the representability of the probability sequence K_3 in a classical event space, it follows that for all triples $(\mathbf{n}, \mathbf{n}', \mathbf{n}'')$ which satisfy (4.55) the probabilities under consideration can be attributed to an object in accordance with quantum mechanics. Hence for experimentally testing the possibility of probability-attribution one has to select triples $(\mathbf{n}, \mathbf{n}', \mathbf{n}'')$ of vectors that violate (4.55a), (4.55b).

It is not hard to determine the validity regions of (4.55a). For given vectors $\mathbf{n}'$ and $\mathbf{n}''$ these relations are fulfilled for all vectors $\mathbf{n}$ *exterior to or on* the double cones $\mathscr{C}_1$ and $\mathscr{C}_2$ formed by the vectors $\mathbf{r}$ satisfying the equations $|\mathbf{r} \cdot (\mathbf{n}' - \mathbf{n}'')| = |\mathbf{n}' \cdot (\mathbf{n}' - \mathbf{n}'')|$ and $|\mathbf{r} \cdot (\mathbf{n}' + \mathbf{n}'')| = |\mathbf{n}' \cdot (\mathbf{n}' + \mathbf{n}'')|$, respectively (Fig. 4.8). Hence, triples $(\mathbf{n}, \mathbf{n}', \mathbf{n}'')$ of vectors that can be used for an experimental test of probability attribution must be chosen such that $\mathbf{n}$ lies in the *interior* of one of the cones $\mathscr{C}_1$ and $\mathscr{C}_2$ [BLM 92].

The size of the invalidity domain of the Bell inequalities (4.55) is again a measure of the strength of the classicality condition (*CC*). Here the *invalidity* domain Δ_{CC} is given by the union of the interiors of the cones $\mathscr{C}_1$ and $\mathscr{C}_2$ in Fig. 4.8. One can compare this domain with the region Δ_{WO} in which the weak objectification (*WO*) condition (4.38) is violated. This condition was discussed for the same EPR experiment in subsection 4.3(d). For given

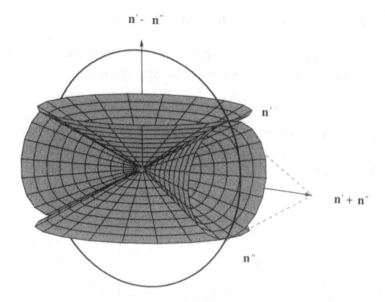

n' - n"

n'

n' + n"

n"

Fig. 4.8 The validity domain of the Bell inequalities is given by those **n** lying outside or on the two double cones.

vectors **n'** and **n"** the weak objectification condition (4.38) can also be written in the equivalent form

$$(\mathbf{n'} \cdot \mathbf{n''}) = (\mathbf{n} \cdot \mathbf{n'})(\mathbf{n} \cdot \mathbf{n''}), \tag{4.56}$$

which is more convenient for the present discussion. It can then be easily seen by straightforward and elementary calculation that all vectors **n** that fulfil eq. (4.56) satisfy also the inequalities (4.55). Hence, these vectors lie in the exterior of or on the double cones $\mathscr{C}_1$ and $\mathscr{C}_2$ (on a one-parameter closed curve).

It is obvious from these considerations that the invalidity region Δ_{CC} of the classicality condition is smaller than and contained in the invalidity region Δ_{WO} of the weak objectification condition. Consequently, the requirement of probability-attribution is weaker than the weak objectification condition. This confirms the result that was obtained in the example with two properties discussed above within the framework of classicality theorem I. It is, however, still an open question whether the probability-attribution postulate is weaker than the requirement of weak objectification also in the general case, i.e., for an arbitrary number of properties.

5

Universality and self-referentiality in quantum mechanics

5.1 Self-referential consistency and inconsistency

5.1(a) Universality implies self-referentiality

In the preceding chapters, we have mentioned on several occasions that there are good reasons to consider quantum mechanics as universally valid. Indeed, during the last 70 years quantum mechanics has not been disproved by a single experiment. In spite of numerous attempts to discover the limits of applicability and validity of this theory, there is no indication that the theory should be improved, extended, or reformulated. Moreover, the formal structure of quantum mechanics is based on very few assumptions, and these do not leave much room for alternative formulations. The most radical attempt to justify quantum mechanics, operational quantum logic, begins with the most general preconditions of a scientific language of physical objects, and derives from these preconditions the logico-algebraic structure of quantum mechanical propositions [Mit 78,86], [Sta 80]. There are strong indications that from these structures (orthomodular lattices, Baer*-semigroups, orthomodular posets, etc.) the full quantum mechanics in Hilbert space can be obtained. Although simple application of Piron's representation theorem [Pir 76] does not lead to the desired result, [Kel 80], [Gro 90], there are new and very promising results [Sol 95] which indicate that the intended goal may well be achieved within the next few years. Together with the experimental confirmation and verification of quantum mechanics, these quantum logical results strongly support the hypothesis that quantum mechanics is indeed universally valid.

Once the universality of the theory is taken for granted, it is obvious that quantum mechanics can also be applied to experimental setups which are used as measuring apparatuses. This means that quantum mechanics not only provides a description of quantum objects the properties of which are

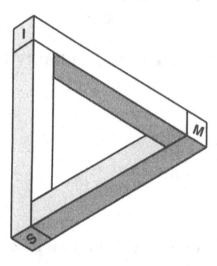

Fig. 5.1 The twofold rôle of the measuring apparatus illustrated by the impossible 'tribar'. Adopted from [Pen 58] and redesigned. S = object system, M = apparatus, I = interpretation.

tested by independent measuring instruments but also governs the measuring processes that are used for its own justification. Einstein's famous statement that 'it is the theory which decides what we can observe' (cf. [Heis 69], p. 92) is obviously extended in the present case: quantum mechanics describes the full measuring process and treats the apparatus as a proper quantum system. This property whereby a theory incorporates the theory of its own measuring instruments was called in the preceding chapters 'semantical completeness'. Hence, *universality* implies *semantical completeness*. It is obvious that the inverse implication does not hold.

As mentioned above, there are good reasons to assume that quantum theory is universally valid. Since universality implies semantical completeness, it also follows that quantum mechanics is a self-referential theory. On the one hand the measuring process serves as a means for verifying and falsifying the theory in question and thus for establishing a semantics of truth for this theory. With respect to this rôle, the measuring process is part of the metatheory and intimately related to the possibilities of interpreting the theory. On the other hand, considered as a quantum mechanical phenomenon, the measuring process is subject to the laws of quantum object theory, and the measuring apparatus is treated as a proper quantum system. This twofold rôle of the measuring process has the consequence that in each case it depends on the context: does the process belong to the object theory or to the metatheory? This ambiguity of the measuring process is illustrated by the 'tribar' in Fig. 5.1, which (vainly) tries to overcome the impossibility of being object (system) and subject (observer-apparatus) at the same time.

The twofold rôle of the apparatus and the measuring process leads to some new questions that relate to the internal consistency of the theory. The interpretation I in Fig. 5.1, which is applied to the object theory and the measuring process, should be *compatible* with the implications of the measuring process for possible interpretations. As shown in the preceding chapters, different parts of interpretation I lead to different answers to the consistency question. In one case, one finds something more than compatibility, a result which was called self-consistency. In another case, one finds that the interpretation and the measuring process are incompatible, a result that indicates a serious internal inconsistency of quantum mechanics.

5.1(b) Self-consistency: The statistical interpretation

The first case mentioned above refers to the *statistical part* of the considered interpretation, which may be either the minimal interpretation I_M or the realistic interpretation I_R. The quantum theory of measurement is not merely compatible with this part of the interpretation, it is even capable of yielding it. This self-consistency of the quantum mechanical formalism was discussed in detail in chapter 3. We add here a few remarks that concern the self-consistency of the theory in the sense of Fig. 5.1.

Universality implies not only semantical completeness but also the applicability of quantum mechanics to any arbitrary quantum system. This means that quantum mechanics can be applied to a single object system $S(\varphi)$ as well as to a compound system $S^{(N)}$ that is composed of N identically prepared systems $S_i(\varphi)$, with $i = 1, 2, \ldots, N$. On the basis of this assumption, one can not only show the compatibility of the interpretation in question with the quantum theory of measurement; in addition, it follows from the applicability of quantum mechanics to the ensemble $S^{(N)}$ that the statistical part of the interpretation I_M, say, can be treated by means of quantum theory. In this way, it can be shown that the probabilistic component of I_M, i.e., the probability reproducibility condition (PR), is a consequence of the probability-free interpretation I_M^0 and the quantum theory of measurement. It is obvious that this internal consistency of quantum mechanics exceeds the mere compatibility of an arbitrary semantically complete system.

Universality implies semantical completeness and semantical completeness implies self-referentiality. In the present case of the statistical interpretation, the self-referential quantum theory proves to be not only consistent in the sense of compatibility; in addition, the statistical part of the interpreta-

tion, i.e., the probability reproducibility conditions (PR) and (SR), can even
be deduced from the probability-free theory. This result leads to an internal structure that was called 'self-referential consistency' and is illustrated
schematically in Figs. 3.1 and 3.2.

5.1(c) Self-referential inconsistencies: Objectification

The second case refers to the objectification of the pointer observable or
of the system observable. The postulates (PO) and (SO) are inevitable
components of the interpretations I_M and I_R, respectively. However, the
quantum theory of measurement is not capable of deriving these important
postulates. Moreover, the objectification postulates are actually incompatible
with the quantum theory of measurement. This result, which was extensively
discussed in chapter 4, obviously violates the compatibility requirement
that should be fulfilled by any self-referential theory. For this reason, the
incompatibilities between quantum theory and the postulates (PO) and (SO)
of the interpretations I_M and I_R respectively were called 'self-referential
inconsistencies'. They are illustrated schematically in Figs. 4.4 and 4.5.

The physical relevance of the nonobjectification theorems becomes obvious
if one recalls the reasoning that they are based on: suppose that the values
of the pointer observable after the measurement are objectively decided
(though subjectively unknown). Then the compound system $S + M$ is (after
the measurement) in a *mixture of states* that can be described by a *mixed state*
$W_{\Psi'}$. However, according to the quantum theory of unitary premeasurements,
the state of the compound system $S + M$ after the premeasurement is given
by the pure state $\Psi' = U(\varphi \otimes \Phi)$, where U is the unitary operator of the
premeasurement and φ and Φ are the preparation states of the system S and
the apparatus M, respectively. Hence, one arrives at an obvious contradiction,
since the system $S + M$ cannot be simultaneously in a mixed state $W_{\Psi'}$ and
a pure state Ψ'.

From a logical point of view, this contradiction between the objectification postulates and the quantum theory of measurement is a serious
self-referential inconsistency. For this reason, the objectification requirements must be rejected. However, there are many physical situations that
are not directly affected by this contradiction. Indeed, the states $W_{\Psi'}$ and
Ψ' are distinguishable only by means of measurement results of an observable $\Theta(S + M)$ of the compound system. If for an observable of this kind
the probability distributions $p(\Psi', \Theta_i)$ and $p(W(\Psi', \Theta_i))$ are different, then
the contradiction becomes obvious and the objectification hypothesis is no
longer tenable. Since for experimental demonstration of the difference be-

tween these probabilities interference experiments with the *compound* system must be performed, which is most *impracticable*, in the majority of realistic physical situations the nonobjectifiability of some object observables is not directly recognizable.

Summarizing these arguments we find that from a logical point of view the nonobjectification theorems show up a very serious self-referential inconsistency of quantum mechanics, which must be resolved in some way. However, for practical purposes the nonobjectification inconsistency is of minor importance since the contradiction is almost unrecognizable. At first glance, the latter remark does not seem to be very relevant for the foundations of quantum physics. It will be shown in section 5.3 that there are situations in which the above-mentioned impracticability proves in fact to be a fundamental impossibility.

5.2 The classical pointer

5.2(a) Back to Bohr: An inconsistency

The nonobjectification theorems that we are considering have serious consequences for the quantum theory of measurement. In the case of repeatable measurements, one should expect the objectification of system values, but this system objectification cannot be achieved by unitary repeatable premeasurements. The realistic calibration postulate can be fulfilled, i.e., an objective value of A prior to the measurement is still objective after a repeatable A-measurement. However, it does not follow from this result that the observable A possesses objective values after the measurement in the general case of an arbitrary preparation. As far as we know, there is no remedy against this negative result, which, however, does not completely invalidate the quantum theory of measurement. One could accept the fact that there is no repeatable measurement at all that leads to objective values of the object observable after the measurement.

In contrast to the possibility of dispensing with system objectification, the objectification of pointer values is inevitable for any measurement. The nonobjectifiability of the pointer observable, as established in (NO) theorem II, plainly contradicts one of the three basic principles of the *minimal interpretation*. For this inconsistency, it does not matter whether one considers repeatable or nonrepeatable measurements. Hence, the theory disproves its own interpretation, making statements that are hard to believe: after the measurement, the macroscopic pointer of the apparatus does not possess an objective value of the pointer observable.

Since this theoretical result is not in agreement with the everyday experience of experimentalists, even if they are experts in quantum physics, one could try to escape the strange consequences of (*NO*) theorem II by arguing the other way round. One could start from the experience that experimental setups provide definite results and thus assume that definite values pertain to the pointer after the measurement. In order to investigate whether this way of reasoning is a promising attempt, we consider some obvious consequences of the assumption mentioned.

If the pointer observable $Z = \sum Z_i P[\Phi_i]$ can be weakly objectified with respect to the mixed state W_M' of the apparatus after the premeasurement, then the observable

$$\mathbf{Z} = \mathbb{1}_S \otimes Z = \sum_i Z_i (\mathbb{1}_S \otimes P[\Phi_i])$$

can be weakly objectified with respect to the compound state

$$\Psi'(S + M) = \sum (\varphi^{a_i}, \varphi)\varphi_i' \otimes \Phi_i$$

after the unitary premeasurement. From this assumption, it follows that the projection operator $P[\Psi']$ is equivalent to the mixed state

$$W'(\Psi', \mathbb{1}_S \otimes Z) = \sum_i (\mathbb{1}_S \otimes P[\Phi_i]) P[\Psi'](\mathbb{1}_S \otimes P[\Phi_i]), \qquad (5.1)$$

which one would obtain by a Lüders measurement of the observable $\mathbf{Z}$. (For the derivation of this result, see subsections 4.2(a) and 4.3(b). The equivalence of $P[\Psi']$ and $W'(\Psi', \mathbf{Z})$ means that for any projection operator $R \in \mathscr{P}(\mathscr{H}_M)$ of the apparatus the equation

$$\mathrm{tr}\{P[\Psi'](P \otimes R)\} = \mathrm{tr}\{W'(\Psi', \mathbf{Z})(P \otimes R)\} \qquad (5.2)$$

holds for all projections $P \in \mathscr{P}(\mathscr{H}_M)$, i.e., the expectation value of the product operator $P \otimes R$ does not depend on whether the compound system is in the pure state Ψ' or in the mixed state $W'(\Psi', \mathbf{Z})$. From eqs. (5.1) and (5.2), one obtains the equation

$$\mathrm{tr}\{P[\Psi'](P \otimes R)\} = \mathrm{tr}\left\{P[\Psi']\left(P \otimes \sum_i P[\Phi_i]RP[\Phi_i]\right)\right\} \qquad (5.3)$$

and thus $[R, Z] = 0$ for any $R \in \mathscr{P}(\mathscr{H}_M)$. Since the commutators $[R, Z]$ vanish for arbitrary R it follows that Z is a classical observable.

At first glance, this result looks somewhat surprising. However, one should remember that the classicality of the measuring apparatus was one of the basic requirements of the *Copenhagen* interpretation. Hence one gets the impression that our way of reasoning has led us back to Bohr's original interpretation. This might be correct, but it turns out that classicality of the pointer observable is not compatible with the other postulates of the minimal interpretation. In order to demonstrate this result, let us assume that Z is a classical observable that commutes with all observables of the apparatus. Then it follows also that the compound observable $\mathbb{1}_M \otimes Z$ commutes with all observables $\Theta(S + M)$ of the compound system $S + M$, in particular with the interaction Hamiltonian $H_{\text{int}}(S + M)$, i.e.,

$$[\mathbb{1}_S \otimes Z, H_{\text{int}}(S + M)] = 0 \tag{5.4}$$

and on account of $U = \exp(-\frac{i}{\hbar} H_{\text{int}} t)$ we obtain

$$[\mathbb{1}_S \otimes Z, U(S + M)] = 0. \tag{5.5}$$

If one calculates the expectation value of Z in the postmeasurement state $\Psi' = U\Psi$, one finds on account of (5.5)

$$(\Psi', Z\Psi') = (U\Psi, ZU\Psi) = (\Psi, U^\dagger Z U\Psi)$$
$$= (\varphi \otimes \Phi, \mathbb{1}_S \otimes Z \varphi \otimes \Phi) = (\varphi, \mathbb{1}_S \varphi)(\Phi, Z\Phi)$$

and thus

$$(\Psi', Z\Psi') = (\Phi, Z\Phi) \tag{5.6}$$

This means that the expectation value of the pointer observable after the premeasurement does not depend at all on the preparation φ of the object system. Consequently, the initial probability distribution $p(\varphi, a_i)$ is no longer reproduced in the statistics of the pointer values. Instead, the statistics of the classical pointer is given by the Gemenge

$$\Gamma_{\text{class}}(Z) = \{|(\Phi, \Phi_i)|^2, Z_i\} \tag{5.7}$$

with weights $w_i = |(\Phi, \Phi_i)|^2$. Consequently, the pointer statistics depends exclusively on the preparation Φ of the pointer and no longer on the preparation φ of the object system. Since this result plainly contradicts the probability reproducibility condition (PR) for the pointer values, it also follows that the assumed pointer objectification (PO) contradicts the condition (PR), i.e.,

$$(PO) \Rightarrow \neg(PR) \tag{5.8}$$

We thus arrive at the important conclusion that the assumption of objective pointer values would contradict the well-confirmed experience that the statistics of pointer values depends significantly on the preparation of the object system.† Summarizing the arguments that follow from the assumed pointer objectification we arrive at the following results.

1. For unitary premeasurements (U), pointer objectification (PO) implies classicality of the pointer observable (CP):

$$(PO) \wedge (U) \Rightarrow (CP) \tag{5.9}$$

and vice versa.

2. If the unitary measurement coupling $U = \exp(-\frac{i}{\hbar}H_{\text{int}}t)$ is generated by an observable H_{int} of the compound system $S + M$, (HO), then it follows from the classicality (CP) that the probability reproducibility condition (PR) is violated, i.e.,

$$(CP) \wedge (HO) \Rightarrow \neg(PR) \tag{5.10}$$

Within the framework of unitary premeasurements, from the relations (5.9) and (5.10) the following conclusions can be drawn. On the one hand, if the probability reproducibility condition (PR) is considered as inevitable, then one must either dispense with classicality and thus with pointer objectification (PO) or accept the strange conclusion that the Hamiltonian of the compound system $S + M$ is not an observable. On the other hand, if one insists on (PO) and (HO) it follows that the probability reproducibility condition (PR) cannot be fulfilled. In the triangle shown in Fig. 5.2, one can satisfy simultaneously in each case only two of three requirements.

5.2(b) Possible solutions

The triangle in Fig. 5.2. shows where one can hope to find a solution of the inconsistencies mentioned. Here we give a brief report of various attempts to find a way out. It should be emphasized that up to now none of these attempts has really been successful. They can only be considered as possible strategies for attacking the measurement problem in the near future. We classify these strategies in two categories, one of which insists on the pointer objectification while the other dispenses with pointer objectification.

---

† A second problem that arises from the assumption of pointer objectification can be mentioned here only briefly, since its mathematical details are beyond the scope of the present book. According to the previous discussion, pointer objectification implies that the pointer observable Z is classical. Furthermore, if Z is a discrete nondegenerate observable it follows that the apparatus M is a discrete classical system. However, a system of this kind cannot be the carrier of Galileo-covariant position and momentum observables [Mack 63], [Pir 76]. Generally, a quantum system is constituted by fundamental Galileo-covariant position and momentum observables. In formal terms, this means that an object quantum system is given by a Galilean system of covariance (a system of *imprimitivity*) [Mit 95], [BGL 95]. Hence, one is led to the conclusion that a discrete classical apparatus M cannot be a quantum system. Obviously, this consequence contradicts the universality of quantum mechanics, which is presupposed throughout the present book.

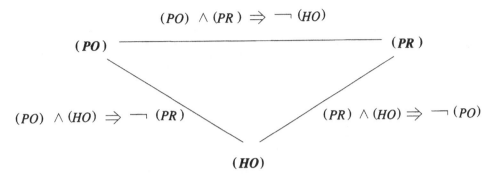

Fig. 5.2 The inconsistency of unitary measurements: each pair of requirements contradicts the third one.

Attempts to preserve pointer objectification

Open systems. The first attempt replaces the isolated quantum system $S + M$ by an open system $S + M$ and an environment E. The new isolated system $S + M + E$ is then governed by a unitary dynamics U, with Hamiltonian H that leads to a nonunitary evolution of the subsystem $S + M$ [Zeh 70]. In this way the environment E could dynamically influence the compound system $S + M$ in such a way that pointer objectification is achieved with a very high degree of accuracy [Zur 82]. The quantum mechanical coherence originally present in the compound system $S + M$ is not really destroyed by the environment but merely displaced into the many degrees of freedom of system E. In particular, if the environment is assumed to possess an infinite number of degrees of freedom, then the induced objectification will even be exact. However, the consideration of infinite systems exceeds the framework of Hilbert-space quantum mechanics.

There are several interesting examples that show how an environment-induced objectification could work. Properties of molecules such as chirality or geometrical structure can be shown to be almost stabilized by an environment of low-energy photons [Prim 83], [Pfei 80], [Ama 88]. In this way, the obvious incompatibility of quantum mechanical predictions and the physico-chemical experience disappears in all practical cases. One could imagine that the mechanism that is demonstrated by these examples works even more for mesoscopic and macroscopic properties. The paradox of Schrödingers's cat can then be resolved by such arguments, since in a realistic environment the cat is objectively dead or alive. Another way to understand the observed stability of molecular states that are not stable in an isolated system consists of incorporating collisions with other molecules. It has been shown by

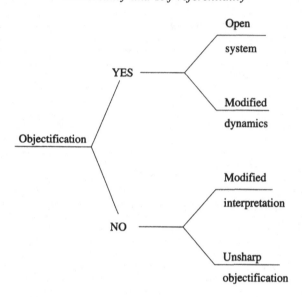

several authors [Sim 78], [Joos 84], [JoZe 85] that collisions of this kind could provide a 'watch dog' effect on the considered systems.

There are two arguments against this way of reasoning. First, the problem of objectification is not really solved. The quantum mechanical coherence of the system $S + M$, which is the reason for the nonobjectification of the pointer, is merely displaced into the many degrees of freedom of the environment E. In addition, this displacement is never complete, and hence only an approximate objectification can be achieved in this way – except when the environment system is infinite. The latter case is, however, beyond the scope of Hilbert-space quantum mechanics. Second, even if this displacement is considered as an approximate but sufficient solution of the objectification problem, this solution is not always applicable. It is obvious that the incorporation of the environment E will lead to a new isolated system $S + M + E$ and thus restore the objectification problem. In most cases, one could of course add a larger environment E' such that $S + M + E + E'$ is the new total system and $S + M + E$ is no longer isolated. The iteration of this procedure is also called the *von Neumann chain*. However, there is a situation in which this iteration has a well-defined limit and the resulting system $S + M$ cannot be embedded in a larger environment. This is the case in quantum cosmology, in which S is the entire universe and the apparatus M is contained in S. On the other hand, a quantum mechanical treatment of the universe is unavoidable for investigating the very early stages of cosmological evolution.

Irrespective of the two objections that we have mentioned there remains a more philosophical dissatisfaction with the environment approach. If the approximate objectification of the pointer is only induced by the environment, one must conclude that the very conditions of quantum physical cognition depend on whether we – or more precisely $S + M$ – are isolated or in a suitable environment E. In an isolated compound system $S + M$, the observer M would then be completely unable to acquire any objective information about his external reality S.

Modified dynamics. The second attempt to preserve pointer objectification consists of a modification of the dynamics. Since the ordinary linear and unitary evolution does not lead to objective pointer values, one could try to dispense with the linearity of the dynamics and to obtain pointer objectification in this way. The first proposal in this direction was made by Wigner [Wig 63], but no explicit formulation of a nonlinear Schrödinger equation was given at that time. The presently best-known, and well-elaborated, example of a nonlinear dynamics is the theory of Ghirardi, Rimini, and Weber [GRW 86]. These authors propose an explicit modification of the well-known unitary dynamics. Since the interference pattern of microscopic quantum systems is a well-established fact, a nonlinear extension of quantum mechanics must fulfill two important requirements. For microscopic systems it should essentially preserve the results of ordinary quantum mechanics, but for macroscopic quantum systems the nonlinearity should provide some damping mechanism that destroys the quantum mechanical coherence in such a way that pointer objectification becomes possible. The theory of Ghirardi, Rimini, and Weber fulfills these requirements. Another general framework for nonlinear evolution in quantum mechanics has been proposed by Weinberg [Wei 89].

In spite of the obvious success of this approach there are several serious objections to it. It has been pointed out by Gisin and by Pearle that a nonlinear term in the dynamical law will presumably lead to the possibility of superluminal signals [Gis 89,90], [Pea 86]. It is not entirely clear whether these results are compatible with the basic requirements of special relativity. Another open problem of the *unified dynamics* is the description of mesoscopic and macroscopic quantum effects. If the nonlinearities provide an overall damping of quantum correlations on the macroscopic level, a theory of this kind will not only lead to dynamical objectification of the pointer but will also destroy any other macroscopic coherence. The well-known effects of superconductivity, quantized flux, macroscopic tunnelling etc. might then not be adequately described by the theory. It is still an open question whether one can reformulate this unified dynamics in such a way that the known macroscopic quantum effects can be preserved together with pointer objectification.

Finally, it should be emphasized that any modified dynamics is at present purely phenomenological and not justified by first principles. A theory like unified dynamics could perhaps be derived from other assumptions as, e.g., the environment approach. However, derivations or justifications of the unified dynamics from more fundamental axioms are presently not yet known.

Attempts without pointer objectification

Modified interpretation. Instead of modifying the dynamics of the quantum mechanical evolution, one could also try to modify the interpretation without changing the object theory. On account of the self-referentiality of quantum mechanics mentioned in subsection 5.1(b), there are of course limits for modifying the semantics. Nevertheless, the self-referential inconsistencies discussed in subsection 5.1(c) are a strong indication of the need to change the interpretation by removing the requirements of objectification.

The many-worlds interpretation first formulated by Everett [Eve 57] and Wheeler [Whe 57] tends in this direction. Since the formalism of quantum mechanics justifies neither system objectification nor pointer objectification, the Everett–Wheeler relative-state interpretation dispenses altogether with the objectification requirement. Instead, it tries to read off the interpretation from the formalism of quantum mechanics in the version of von Neumann, at least for discrete observables and repeatable measurements. For this reason, the many-worlds interpretation should not be considered as an alternative interpretation (with respect to the minimal interpretation) but rather as a precise formulation of the physical situation that is described by quantum mechanics. Since no additional hypothesis in incorporated into the interpretation, the missing objectification leads to a concept of reality that corresponds to many alternative worlds.

With respect to the problem of objectification, the many-worlds interpretation does not provide a resolution. Rather, it shows that quantum mechanics as it stands is not a theory which refers to an individual system and its properties. Instead, it seems to be a theory which has as its referent an infinite ensemble of many distinct alternative worlds. Since such a theory was neither intended by the founders of quantum mechanics nor corresponds to our present expectations, one gets the impression that the formulation of quantum mechanics ought to be changed in some way. It is obvious that this conclusion leads us back to the attempts to modify the dynamics mentioned above.

Unsharp objectification. The last way out of the disaster of objectification assumes that quantum mechanics in Hilbert space is universally valid and applicable to the *one* world in which we are living. Since under these conditions objectification cannot be achieved, it follows that there is no pointer objectification after the measuring process. At first glance, this conclusion seems to contradict the overall experience of experimental physicists. However, it is still possible that exact and sharp pointer objectification is an illusion and that the results obtained by experimentalists are objectified only in an unsharp sense, such that the contradiction with the formalism disappears but the results are objective for all practical purposes.

According to the results of chapter 4, which are summarized in subsection 5.1(c), for ordinary system observables corresponding to self-adjoint operators A and pointer observables Z of the same kind, objectification of the pointer values after the measuring process cannot be achieved. However, one might guess that by means of an appropriate relaxation of the concept of the measured observable the problems we are considering would disappear. The generalization that we have in mind here consists of replacing the self-adjoint operators that correspond to projection-valued measures by 'positive operator-valued' (POV) measures. These unsharp observables are not the subject of the present book and will be mentioned here only briefly. For details, we refer to the literature, in particular to [BLM 96] and [BGL 95]. Here we confine our considerations to a few remarks which show that unsharp observables could probably solve the measurement problem.

In contrast to ordinary self-adjoint operator observables, for unsharp noncommuting observables there exist joint measurements that allow for a simultaneous unsharp observation of two incommensurable unsharp properties [MPS 87], [BuS 89]. Another indication that unsharp observables might be a useful tool for solving the objectification problem is the observation that the Bell inequalities are satisfied within quantum mechanics if sufficiently unsharp observables are considered [Bus 85], [Fine 82]. Hence, there could be some hope that the nonobjectification theorems discussed in chapter 4 lose their relevance for POV observables.

This is, however, not the case. According to a recent result [BuShi 96], [Bus 96], also for unsharp system observables in the sense of POV measures a nonobjectification theorem, which is similar to that for ordinary sharp observables, can be proved. The theorem states that there is no measurement of a given unsharp observable that provides objectification for all initial states of the object system and the apparatus. Hence, POV observables of the system do not lead to an appropriate concept of unsharp objectification even if these observables are useful for other reasons. Moreover, those unsharp pointer observables that assume definite pointer values in the final component states of the apparatus must also be excluded [Bus 96].

The only way to realize the idea of unsharp objectification in accordance with quantum mechanics that is still possible consists of assuming that the pointer observable Z itself is a genuinely unsharp observable. This means that the post-measurement apparatus mixed state W'_M can be interpreted as a Gemenge of final component states that do not correspond to definite pointer values but rather to *almost real* properties. At first glance, this solution looks reasonable. However, even if it works one should keep in mind its far-reaching consequences: *one has to give up the idea of an objective reality.* Indeed, not only microscopic quantum systems but also macroscopic instruments and pointers would be in a state of objective undecidedness that is expressed by the genuine unsharpness of the pointer observables. Even if the degree of objectification is very high in all practical cases the observations will always contain a finite amount of nonobjectivity. Schrödinger's cat, considered as a macroscopic pointer of an apparatus, is never dead or alive in a rigorous sense in its final state. The use of genuinely unsharp pointer observables can presumably help to eliminate the inconsistencies in the formalism of quantum mechanics, but it must not be considered as a means to restore objectivity and reality in the physical world (see Fig. 5.3).

5.3 The internal observer

5.3(a) Measurements from inside

The quantum theory of measurement that was discussed in the preceding chapters considers two distinct proper quantum systems, the object system S and the apparatus M. Usually, the apparatus is a large macroscopic system whereas the object system is an atom or a molecule, etc. However, in the formalism both systems are treated as proper quantum objects whose size has no influence on the formal procedure. Prior to the measuring process, the systems S and M are isolated and may be also separated in space. Hence, the state of the formal composite system $S + M$ is given by the tensor product state $\Psi(S + M) = \varphi(S) \otimes \Phi(M)$, where φ and Φ are the initial pure states of S and M, respectively. The unitary operator U of the premeasurement acts on the compound state Ψ and leads after an appropriate time to a highly correlated state $\Psi' = U\Psi$. Since Ψ' is an entangled state, the reduced mixed states W'_S and W'_M of S and M, respectively, after the premeasurement provide only restricted information about the composite system, since they do not contain the correlations between S and M. Here we consider a

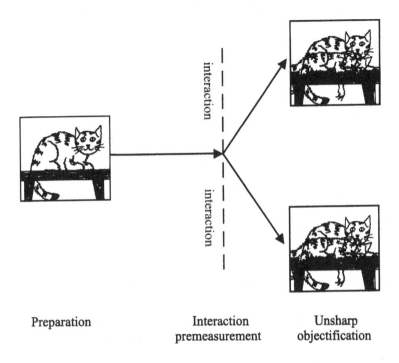

Preparation · Interaction premeasurement · Unsharp objectification

Fig. 5.3 Schrödinger's cat considered as the pointer of an apparatus. If the property of being dead or alive is a genuinely unsharp pointer observable, then the cat will be neither dead nor alive after the measurement.

different situation in which the apparatus M is contained in the system S as a proper subsystem, i.e., $M \subset S$.

At first glance, this arrangement looks rather artificial. There are, however, two reasons for discussing this problem. First, in quantum cosmology the object system S is given by the entire universe. Under these conditions, it is obvious that the measuring apparatus M that is used for measuring observables of S must be contained in S as a proper subsystem. Second, measurements from inside have a theoretical interest since they make it possible to derive some general results that prove to be useful for other problems, e.g., for pointer objectification. In order to make the situation definite, we shall not distinguish here the apparatus and the observer but rather assume that *the measuring apparatus plus observer* corresponds to the system M.

Let S be the object system and M be the measuring apparatus plus observer, which is contained in S as a proper subsystem, i.e., $M \subset S$. The object system S is then composed of subsystems M and R, where $R = S - M$ is the residue, and thus $S = M + R$ (Fig. 5.4).

object system S

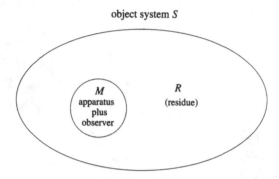

Fig. 5.4 The internal 'apparatus plus observer', M, is a proper subsystem of the object system S, which is measured from inside. Hence $M \subset S$, $R \subset S$, and $M + R = S$.

If the apparatus M is used to measure the values A_i of an observable A of the object system S, then one finds some new restrictions that have at first glance nothing to do with the well-known restrictions of quantum measurements, which come from mutual incommensurabilities. Indeed, we are confronted here with a purely logical argument, which will, however, be formulated here in terms of quantum mechanics. For a more general formulation of these results, we refer to the work of Breuer [Breu 94, 96].

5.3(b) Limits of recognizability

Let Z be again a discrete nondegenerate pointer observable with values Z_i and states Φ_i, and A be an object observable with values A_i and states φ_i. Generally, a measurement of A should lead to a definite pointer value Z_i which indicates the measurement result $A_i = f(Z_i)$, where f is the pointer function. A measurement from inside is characterized by some peculiarities which can be expressed by two additional requirements.

I *Condition of proper inclusion.* If M is properly included in S, then the object system S has strictly more degrees of freedom than the apparatus M. This means that there are pairs of different S-states (W, W') with $W' \neq W$ that have the same reduced state $W_M = W'_M$ in M.

II *Condition of self-referential consistency.* If a pointer value Z with state Φ indicates the object value $A = f(Z)$ with state W, then the reduction of the object state with respect to the subsystem M should reproduce the apparatus state Φ.

This consistency condition is first of all a purely logical requirement that guarantees that the twofold rôle of the measuring apparatus does not lead to inconsistencies. In quantum mechanics, this condition is fulfilled in the following sense. If after the measurement the pointer has the value Z_i with the state Φ_i, indicating the object value $A_i = f(Z_i)$ and the object state φ_i, then the state φ_i of the object system is given by $\varphi_i = \Phi_i \otimes \rho_i$ where ρ_i is the pure state of the residue R. It is then obvious that the reduction of the S-state φ_i to the apparatus is given by the state Φ_i. The argument that is used here is based on some well-known properties of the tensor product (cf. e.g. [Jau 68] pp. 179–82 and subsection 2.2(a) of the present text).

We say that a state W of the object system is *uniquely measurable* by a given measurement scheme $\mathcal{M}$ (consisting of observables Z and A, the pointer function f, etc.; cf. chapter 2) if there is a set of apparatus states that refer *uniquely* to the object state W. Two states W and W' of the object system are said to be distinguishable by a measuring scheme if there are two sets of apparatus states, one referring to W and not to W' and the other one referring to W' and not to W. It is an interesting result that the unique measurability and the distinguishability are strongly restricted by the above conditions I and II on a measurement from inside.

Suppose that by means of a certain measurement scheme $\mathcal{M}$ all states of the object system are uniquely measurable. According to the proper inclusion condition I, there are at least two different S-states W, W' with the same reduced state $W_M = W'_M$ in M. If one applies the consistency condition II to the S-state W, say, then the pointer state indicating W must agree with W_M. Since the same argument applies to W' (saying that the indicating pointer state reads $W'_M = W_M$), it follows that the two states W and W' are not uniquely measurable. Hence, we arrive at the following result:

Lemma 5.1 *The conditions of proper inclusion and of self-referential consistency imply that not all states of the object system can be measured uniquely by an internal apparatus plus observer.*

The second of the problems we are considering concerns the distinguishability of two system states W and W'. Consider two states W, W' with the same reduced state $W_M = W'_M$, the existence of which is guaranteed by the proper inclusion condition I. According to the consistency condition II, the apparatus states referring to W and W' are given by the same state $W_M = W'_M$. Hence, the object states W and W' cannot be distinguished by a measurement from inside using the apparatus M. We thus arrive at the second result:

Lemma 5.2 *The conditions of proper inclusion and of self-referential consistency imply that two object states W and W' with the same reduced state $W_M = W'_M$ in M cannot be distinguished by the internal apparatus plus observer.*

Lemmas 5.1 and 5.2 lead to interesting applications, one of which will be discussed in the next subsection.

5.3(c) Subjective objectification

Let us consider again the situation that is described by Fig. 5.4. The apparatus M is a proper subsystem of the system S, which is composed of M and the residue R. Here, however we shall discuss this arrangement from a slightly different angle. The apparatus is used for measuring states and observables of the residue system R. In this case, we can apply the well-established quantum theory of measurement, but using the same notation as in the preceding subsections 5.3(a),(b). The initial state of R will be denoted by ρ and the states and values of the measured R-observable r by ρ_i and r_i, respectively. A unitary repeatable measurement of the observable $r = \sum r_i P[\rho_i]$ is then described by the calibration condition

$$U(\rho_i \otimes \Phi_i) = \rho_i \otimes \Phi_i$$

and the premeasurement $\varphi = \rho \otimes \Phi \to U(\varphi)$ with

$$U(\rho \otimes \Phi) = \sum (\rho_i, \rho)\rho_i \otimes \Phi_i =: \varphi'.$$

If the composite system $S = M+R$ is in the pure state φ', then the subsystems M and R are in the reduced mixed states

$$W'_M = \text{tr}_R\{P[\varphi']\} = \sum |(\rho_i, \rho)|^2 P[\Phi_i],$$
$$W'_R = \text{tr}_M\{P[\varphi']\} = \sum |(\rho_i, \rho)|^2 P[\rho_i].$$

According to the nonobjectification theorems of section 4.3, the mixed states W'_M and W'_R do not admit an ignorance interpretation. This means that it is not possible to assume that the system R, say, possesses objectively a value r_i with state ρ_i but that this result is subjectively unknown to the observer. The argument that justifies these nonobjectification results should be briefly recalled, as follows.

If the state W'_R, say, were to admit an ignorance interpretation, then the composite system $M + R = S$ would be in the mixed state $W_{\varphi'} := \sum |(\rho_i, \rho)|^2 P[\rho_i \otimes \Phi_i]$ and not in the pure state φ' that follows from the quantum theory of unitary premeasurement. The two S-states $P[\varphi']$ and $W_{\varphi'}$ are different, but they have the same reduced state in M:

$$\text{red}_M W_{\varphi'} = \text{red}_M P[\varphi'] = W_M'.$$

In order to demonstrate that the states $P[\varphi']$ and $W_{\varphi'}$ are in fact different, one has to determine the probability distributions of a convenient observable B of the compound system $S+M$ in the two considered states. The difference between the two states is then given by the interference terms, which are directly measurable.

Since the two states $P[\varphi']$ and $W_{\varphi'}$ of S are different but with the same reduced state W_M' in M the premise of lemma 5.2 is fulfilled, and one can apply it to the present case. In this way, it follows from lemma 5.2 that the two states $P[\varphi']$ and $W_{\varphi'}$ cannot be distinguished by the internal apparatus plus observer. This means that the internal observer M cannot distinguish the states $P[\varphi']$ and $W_{\varphi'}$ by a measurement from inside. Consequently, for the internal observer there is no reason to reject the ignorance interpretation of the mixed states W_M' and W_R'. The observer in M is not in a position to disprove the objectification hypothesis with respect to the considered mixed states by any realizable measurement. Hence, the objectification of the measurement results and of the pointer values is no longer prohibited – at least for the internal observer M. He is allowed to objectify the pointer values and the results r_i, since he is subjectively not able to disprove this assumption. However, it is obvious that this *subjective objectification* is nothing but a misleading illusion.

In chapter 4, the objectification hypothesis in the reduced mixed states W_M' and W_S' was disproved by the existence of an interference pattern for some observable of the compound system. Hence, it is already known that a refutation of the objectification assumption of pointer values is possible by means of observables of the compound system. In addition to this result, it follows from lemma 5.2 that a refutation of the ignorance interpretation of mixed states can *only* be performed by measurements of the compound system. For this reason, the internal observer may always assume that the pointer values are objectified after the measuring process. He will never be able to disprove his own assumption by experimental evidence.

In subsection 5.1(c), it was already mentioned that the refutation of the objectification hypothesis is of interest from a fundamental point of view, but that the contradiction between objectification and quantum theory is almost unrecognizable. The reason is that the observables of the compound system that are needed for a refutation are not available in the majority of practical situations. The present discussion of measurements from inside has shown that in addition to the practical impossibility of recognizing the contradiction there are situations in which this impracticability proves to be a fundamental impossibility: the internal observer can never observe the contradiction between quantum mechanics and the objectification hypothesis.

5.4 Incompleteness

5.4(a) Universality implies incompleteness

The assumption of the universal validity of quantum mechanics implies that this theory can be applied not only to the usual microscopic object systems such as atoms and molecules but also to the macroscopic measuring apparatus used for the experimental justification or refutation of quantum mechanical propositions. This universal applicability provides the *semantical completeness* discussed in subsection 5.1(a). Metatheoretical statements can be reformulated in terms of object theory and proved or disproved as propositions of object quantum mechanics. The *interface* between object theory and the metatheory is given by the measuring process, the twofold rôle of which was discussed in 5.1(a). As an illustration of these remarks, consider an object system S in the state φ and an observable A with values A_i. The object-theoretical proposition $A_i(S, \varphi)$: 'system $S(\varphi)$ possesses the value A_i of A' can be verified by a measuring process. Hence, the metaproposition $\vdash A_i(S, \varphi)$: '$A_i(S, \varphi)$ is true' can be expressed in terms of object theory by making use of the quantum theory of measurement by $U_A(\varphi \otimes \Phi) = \varphi \otimes \Phi_i$. Here U_A is the unitary measuring operator for the observable A and Φ_i is the pointer state that indicates the result A_i.

In the preceding chapters, we considered several more complicated metapropositions that can be translated into object theory and proved or disproved as object-theoretical quantum mechanical statements. In this way, the probability reproducibility conditions (PR) and (SR) were proved, and the objectification conditions (PO) and (SO) were disproved. In both cases, the procedure was the same: a metatheoretical proposition (PR) or (PO) is reformulated in terms of object theory and then proved or disproved, exclusively by means of object theory. The possibility of transferring a metatheoretical concept into object theory is, in the present case, given by the quantum theory of measurement and based on the assumption that quantum mechanics is universally valid.

This situation has some similarity with the metamathematical problems studied by Gödel [Göd 31]. Gödel investigated a formal system T that fulfills some standard requirements. In particular, the formal system must be rich enough to admit a formulation of the arithmetic of natural numbers. The formal system T used by Gödel is not semantically complete in the described sense since it does not contain expressions referring explicitly to metatheoretical propositions. However, since arithmetic can be incorporated into the system T, the metapropositions can be reformulated by means of the formal system T by making use of Gödel numbers (or other numbering procedures). But these reformulated propositions are not always provable, even if they can be seen to be true otherwise. This is the content of Gödel's incompleteness theorem, which was demonstrated by Gödel for special self-referential propositions that state their own unprovability. Hence, the formal system considered by Gödel does not permit the proof of all true metapropositions.

5.4(b) Relations to Gödel's theorem

As mentioned above, the universality of quantum mechanics implies its semantical completeness: any metatheoretical proposition can be reformulated in terms of object quantum mechanics by making use of the theory of the measuring process. Hence metatheoretical statements play a twofold rôle. They are elements of the semantics that provides relations between observations and theoretical terms, and they are object-theoretical propositions that follow from the quantum theory of measurement. For this reason, one should expect to find here too object-theoretical propositions that are self-referential in the described sense. On account of the analogy with Gödel's work, it could happen that some of these self-referential propositions are not provable within object quantum theory, even if they can be seen to be true otherwise. Hence one should check whether the metatheoretical results of chapters 3 and 4 are invalidated in any sense by Gödel's theorem.

In spite of the obvious analogy, there are important differences between Gödel's investigations and the situation in quantum theory. The metapropositions that have been proved or disproved here by means of object quantum theory are not self-referential and paradoxical in the sense of Gödel's formula. They are not Gödel-type propositions. For example the metaproposition $\neg(PO)$ proved in chapter 4 (the $(NO)_M$ theorem) does not state its own nonobjectifiability but rather the nonobjectification of the pointer observable. These arguments show that the proofs of metapropositions in chapters 3 and 4 which were performed within the framework of quantum theory are not necessarily in conflict with Gödel's result. Indeed, Gödel's theorem does not exclude the possibility of proving or disproving some special propositions of the metatheory by means of the object theory, which fulfills the requirements mentioned above.

Nevertheless, Gödel's theorem has the consequence for an object theory of this kind that there are true metatheoretical propositions that cannot be proved within the object theory. Quantum mechanics is a theory that presumably fulfills the requirements of Gödel's theorem and that allows for the reformulation of metapropositions in terms of object theory by means of quantum measurements. Hence, one must be prepared to discover true quantum mechanical metapropositions whose object-theoretical reformulation cannot be proven within the object theory. Obviously, these undecidable propositions do not correspond to decidable experimental situations. It is an open question what such Gödel-type propositions in quantum mechanics look like. At first glance, one could think that propositions of this kind are – as Gödel's famous example – extremely complicated and uninteresting for all practical purposes. However, there are indications [Mun 93] that simple and tractable statements of the Gödel type can be constructed in quantum theory. Hence, it could happen that one day Gödel incompleteness phenomena will actually affect the experimental as well as the theoretical quantum physicist.

Appendices

Using the spectral decomposition of F_k^N one obtains, for the trace

$$T_k^{N,M} := \mathrm{tr}\left\{(W_M')^N (F_k^N - p(\varphi, a_k))^2\right\}$$

in the basis Φ_l^N,

$$T_k^{N,M} = \sum_l \left(\Phi_l^N, (W_M')^N (F_k^N - p(\varphi, a_k))^2 \Phi_l^N\right)$$

$$= \sum_l (\Phi_l^N, (W')^N \Phi_l^N)(f^N(k,l) - p(\varphi, a_k))^2. \qquad (A1.1)$$

Lemma A.1

$$(\Phi_l^N, (W_M')^N \Phi_l^N) = \prod_{i=1}^N p^{(i)}(\varphi, a_{l_i}) =: p_{\{l\}}$$

with $p_{l_i}^{(i)} := p^{(i)}(\varphi, a_{l_i}) = (\varphi_{l_i}^{(i)}, P[\varphi^{(i)}]\varphi_{l_i}^{(i)})$.

Proof of Lemma A.1. According to the tensor product rules one obtains

$$(\Phi_l^N, (W_M')^N \Phi_l^N) = \prod_{i=1}^N (\Phi_{l_i}, W_M'^{(i)} \Phi_{l_i}) =: \prod w_{l_i}$$

where the upper index of $W_M'^{(i)}$ indicates the ith measurement. Since

$$w_{l_i} = \sum_k p^{(i)}(\varphi, a_k)(\Phi_{l_i}, P[\Phi_k]\Phi_{l_i}) = \sum_k p^{(i)}(\varphi, a_k)\delta_{k,l_i} = p^{(i)}(\varphi, a_k),$$

it follows that

$$(\Phi_l^N, (W_M')^N \Phi_l^N) = \prod_{i=1}^N p^{(i)}(\varphi, a_{l_i}) = p_{\{l\}}. \qquad \blacksquare$$

By means of this lemma one obtains

$$T_k^{N,M} = \sum_l (f^N(k,l) - p(\varphi, a_k))^2 p_{\{l\}}. \tag{A1.2}$$

For the evaluation of this sum one needs

Lemma A.2

$$f^N(k) := \sum_l f^N(k,l) p_{\{l\}} = p(\varphi, a_k).$$

Proof of Lemma A.2. Using $f^N(k,l) = \frac{1}{N} \sum_{i=1}^N \delta_{l_i,k}$ one gets

$$f^N(k) = \frac{1}{N} \sum_l (\delta_{l_1,k} + \cdots + \delta_{l_N,k}) p_{l_1}^{(1)} p_{l_2}^{(2)} \cdots p_{l_N}^{(N)}$$

$$= \frac{1}{N} (p_k^{(1)} + p_k^{(2)} + \cdots + p_k^{(N)}) = p(\varphi, a_k). \qquad \blacksquare$$

By means of this result one can prove

Lemma A.3

$$\sum_l (f^N(k,l) - p(\varphi, a_k))^2 p_{\{l\}} = \frac{1}{N} p(\varphi, a_k)(1 - p(\varphi, a_k)).$$

Using this lemma one obtains from eq. *(A1.2)*

$$T_k^{N,M} = \frac{1}{N} p(\varphi, a_k)(1 - p(\varphi, a_k))$$

and hence the desired result

$$\lim_{N\to\infty} T_k^{N,M} = \lim_{N\to\infty} \frac{1}{N} p(\varphi, a_k)(1 - p(\varphi, a_k)) = 0.$$

Proof of Lemma A.3. From the equation (A1.2) one obtains by means of lemma A.2

$$T_k^{N,M} = \sum_l \frac{1}{N^2} \left(\sum_{n=1}^N \delta_{l_n,k} \right)^2 p_{\{l\}} - p(\varphi, a_k)^2 \tag{A1.3}$$

and using again lemma A.2

$$T_k^{N,M} = \frac{1}{N^2} \left\{ \sum_{l'} \left(\sum_{n=1}^{N-1} \delta_{l'_n,k} \right)^2 p_{\{l'\}} + 2(N-1)p(\varphi, a_k)^2 - p(\varphi, a_k) \right\} - p(\varphi, a_k)^2 \tag{A1.4}$$

where the sum runs over all index sequences $l' = \{l_1, l_2, \ldots, l_{N-1}\}$. One can now make use of some complete induction. Indeed, for $N = 2$ one finds immediately

$$T_k^{2,M} = \tfrac{1}{2} p(\varphi, a_k)(1 - p(\varphi, a_k))$$

in accordance with lemma A.3. Furthermore by presupposing the $(N-1)$th relation

$$T_k^{N-1,M} = \frac{1}{(N-1)^2} \sum_{l'} \Big(\sum_{n=1}^{N-1} \delta_{l_n,k} \Big)^2 p_{\{l'\}} - p(\varphi, a_k)^2$$

$$= \frac{1}{(N-1)} p(\varphi, a_k)(1 - p(\varphi, a_k))$$

one obtains by means of eq. (A1.4)

$$T_k^{N,M} = \frac{1}{N} p(\varphi, a_k)(1 - p(\varphi, a_k))$$

completing the induction. ∎

Appendix 2: Contribution of non-first-random sequences

For the scalar product $(\chi_\varepsilon^N, \chi_\varepsilon^N)$ of

$$\chi_\varepsilon^N = \sum_l c_{\{l\}} (\varphi)_l^N \otimes \Phi_l^N - \sum_{l, \delta < \varepsilon} c_{\{l\}} (\varphi)_l^N \otimes \Phi_l^N$$

one obtains

$$(\chi_\varepsilon^N, \chi_\varepsilon^N) = \sum_l p_{\{l\}} - 2 \sum_{l, \delta < \varepsilon} p_{\{l\}} + \sum_{l, \delta < \varepsilon} p_{\{l\}} = \sum_{l, \delta \geq \varepsilon} p_{\{l\}}.$$

Furthermore one gets

$$\sum_{l, \delta \geq \varepsilon} p_{\{l\}} \leq \sum_{l, \delta \geq \varepsilon} \frac{\delta}{\varepsilon} p_{\{l\}} = \sum_l \frac{\delta}{\varepsilon} p_{\{l\}} - \sum_{l, \delta < \varepsilon} \frac{\delta}{\varepsilon} p_{\{l\}} \leq \sum_l \frac{\delta}{\varepsilon} p_{\{l\}}.$$

On account of

$$\sum_l p_{\{l\}} \delta(l) = \frac{1}{N} \sum_k p(\varphi, a_k)(1 - p(\varphi, a_k))$$

it follows that

$$\sum_{l, \delta \geq \varepsilon} p_{\{l\}} \leq \frac{1}{\varepsilon} \sum_l \delta(l) p_{\{l\}} = \frac{1}{\varepsilon N} \sum_k p(\varphi, a_k)(1 - p(\varphi, a_k))$$

and thus

$$(\chi_\varepsilon^N, \chi_\varepsilon^N) \le \frac{1}{\varepsilon N} \sum_k p(\varphi, a_k)(1 - p(\varphi, a_k)) \le \frac{1}{\varepsilon N}.$$ ∎

Appendix 3: Proof of probability theorem II

The proof is similar to that for probability theorem I. Using the spectral decomposition of f_k^N it follows for the trace

$$T_k^{N,S} := \text{tr}\{(W_S')^N (f_k^N - p(\varphi, a_k))^2\}$$

calculated in the basis $(\varphi)_l^N$ that

$$T_k^{N,S} := \sum_l ((\varphi)_l^N, (W_S')^N (\varphi)_l^N)(f^N(k,l) - p(\varphi, a_k))^2.$$

Analogously to lemma A.1 one finds, in case of the post-measurement state $W_S' = \sum_k p(\varphi, a_k) P[\varphi^{a_k}]$,

$$\left((\varphi)_l^N, (W_S')^N (\varphi)_l^N\right) = \prod_{i=1}^N p^{(i)}(\varphi, a_i) = p_{\{l\}}$$

and hence

$$T_k^{N,S} = \sum_l (f^N(k,l) - p(\varphi, a_k))^2 p_{\{l\}}.$$

By means of lemma A.3 it follows again that

$$T_k^{N,S} = \frac{1}{N} p(\varphi, a_k)(1 - p(\varphi, a_k))$$

and thus one obtains the desired result

$$\lim_{N \to \infty} T_k^{N,S} = 0.$$ ∎

Appendix 4: Proof of lemma 3.1

The lemma is stated as

$$\Delta^2(N, \xi) = \frac{N-1}{N}\{(\varphi, A\varphi) - \xi\}^2 + \frac{1}{N}(\varphi, (A - \xi)^2\varphi).$$

According to the definitions

$$\bar{A}^N = \frac{1}{N} \sum_{i=1}^N A^{(i)}, \quad (\varphi)^N = \bigotimes_{i=1}^N \varphi^{(i)}$$

we get

$$\bar{A}^N(\varphi)^N - \xi(\varphi)^N = \frac{1}{N} \sum_i^N \varphi_i^N (A^{(i)} - \xi)\varphi^{(i)}$$

with $\varphi_i^N := \overset{N}{\underset{k \neq i}{\otimes}} \varphi^{(k)}$. By means of the definition

$$\Delta^2(N, \xi) = \|A^{(N)}(\varphi)^N - \xi(\varphi)^N\|$$

it follows that

$$\Delta^2(N, \xi) = \frac{1}{N^2} \sum_{i,l} \left(\varphi^{(i)}(A^{(i)} - \xi)\varphi_i^N, \varphi_l^N(A^{(l)} - \xi)\varphi^{(l)} \right)$$

$$= \frac{1}{N^2} \sum_{i,l} \left(\varphi^{(i)}, (A^{(i)} - \xi)\varphi^{(i)} \right) \left(\varphi^{(l)}, (A^{(l)} - \xi)\varphi^{(l)} \right)$$

$$= \frac{1}{N^2} \sum_{i \neq l} \left(\varphi^{(i)}, (A^{(i)} - \xi)\varphi^{(i)} \right) \left(\varphi^{(l)}, (A^{(l)} - \xi)\varphi^{(l)} \right)$$

$$+ \frac{1}{N^2} \sum_{i=l} \left(\varphi^{(i)}, (A^{(i)} - \xi)\varphi^{(i)} \right)$$

The first sum consists of $N(N-1)$ terms of equal value, the second sum consists of N equal terms. Hence it follows that

$$\Delta^2 = \frac{N-1}{N}(\varphi, (A - \xi)\varphi)^2 + \frac{1}{N}(\varphi, (A - \xi)^2\varphi) \qquad \blacksquare$$

Appendix 5: Probability theorem III implies probability theorem IV

It follows from probability theorem III, that

$$\lim_{N \to \infty} \text{tr}\left\{ P[(\varphi)^N](\bar{A}^N - (\varphi, A\varphi))^2 \right\} = 0 \qquad (A5.1)$$

holds.

Replacing the operator A by one of its spectral components $P[\varphi^{a_k}]$ one obtains

$$\lim_{N \to \infty} \text{tr}\left\{ P[(\varphi)^N](\bar{P}_k^N - p(\varphi, a_k))^2 \right\} \qquad (A5.2)$$

with $\bar{P}_k^N := \frac{1}{N} \sum_{i=1}^N P[\varphi_{(i)}^{a_k}]$ and $p(\varphi, a_k) = (\varphi, P[\varphi^{a_k}]\varphi)$. With a slight change in notation, $\varphi_k^{(i)} := \varphi_{(i)}^{a_k}$, we consider the operator $\bar{P}_k^N = \frac{1}{N} \sum_{i=1}^N P[\varphi_k^{(i)}]$ and apply it to the states

$$\varphi_l^N := \varphi_{l_1}^{(1)} \otimes \varphi_{l_2}^{(2)} \otimes \cdots \otimes \varphi_{l_N}^{(N)}, \qquad l = \{l_1, l_2, \ldots, l_N\}.$$

On account of $P[\varphi_k^{(i)}]\varphi_l^N = \delta_{k,l_i}\varphi_l^N$ one gets

$$\bar{P}_k^N \varphi_l^N = \frac{1}{N}\sum_{i=1}^{N} P[\varphi_k^{(i)}]\varphi_l^N = \frac{1}{N}\sum_{i=1}^{N}\delta_{k,l_i}\varphi_l^N = f^N(k,l)\varphi_l^N,$$

where $f^N(k,l)$ is the relative frequency of the value k in the index sequence l. Hence we have the operator equation

$$\bar{P}_k^N \varphi_l^N = f^N(k,l)\varphi_l^N,$$

which shows that $\bar{P}_k^N$ agrees with the operator f_k^N for the relative frequency of value k. Equation (A5.2) can thus be written as

$$\lim_{N\to\infty} \mathrm{tr}\left\{ P[(\varphi)^N](f_k^N - p(\varphi,a_k))^2 \right\} = 0 \qquad (A5.3)$$

in agreement with probability theorem IV.   ∎

Appendix 6: Proof of classicality theorem I

1. Proof that conditions (4.42) are necessary. For any two elements $x, y \in L_B^{(2)}$, $x \le y$ implies $\neg y \le \neg x$ and thus $p(x \vee \neg y) = p(x) + 1 - p(y)$. Hence, we obtain $p(x) - p(y) = p(x \vee \neg y) - 1 \le 0$, or $p(x) \le p(y)$. By means of this result we find that $a \wedge b \le a$ implies $p(a,b) := p(a \wedge b) \le p(a)$ and $a \wedge b \le b$ implies $p(a,b) := p(a \wedge b) \le p(b)$. From these results we obtain the first two lines of (4.42).

From $a \vee b = \neg(\neg a \wedge \neg b)$ it follows that

$$p(a \vee b) = 1 - p(\neg a \wedge \neg b) \qquad (A6.1)$$

and from

$$\neg a = (\neg a \wedge b) \vee (\neg a \wedge \neg b)$$

and

$$\neg a \wedge b \le \neg(\neg a \wedge \neg b)$$

we obtain

$$p(\neg a) = p(\neg a \wedge b) + p(\neg a \wedge \neg b). \qquad (A6.2)$$

Inserting (A6.2) into (A6.1) we obtain

$$p(a \vee b) = p(a) + p(\neg a \wedge b). \qquad (A6.3)$$

Furthermore from

$$b = (b \wedge a) \vee (b \wedge \neg a)$$

it follows that

$$p(b \wedge \neg a) = p(b) - p(a \wedge b).$$

Inserting this expression into (A6.3), we obtain

$$p(a) + p(b) - p(a \wedge b) = p(a \vee b) \leq 1$$

completing the proof of eqs. (4.42). ∎

This proof illustrates the meaning of the classicality conditions (4.42). If the probabilities of the correlation sequence K_2, (4.41), pertain to a physical system as probabilities of its classical properties, then the properties fulfill the laws of classical propositional logic and the probabilities are subject to the Kolmogorov axioms. From these premises the classicality conditions can easily be derived. Hence probability attribution is understood here always in connection with Boolean propositional logic and Kolmogorovian probability theory.

2. Proof that conditions (4.42) are sufficient. According to Pitovsky [Pit 89], there is a very intuitive geometrical interpretation of the classicality conditions (4.42) that illustrates the sense in which the classicality conditions are also *sufficient* for probability attribution. Consider a three-dimensional Euclidean space $\mathbb{R}^3$ with orthogonal axes p_1, p_2, p_{12}. All vectors $\mathbf{p} = (p_1, p_2, p_{12})$ whose components satisfy the inequalities (4.42) are elements of the closed convex polytope $c(2)$ with vertices given by the vectors (Fig. A6.1)

$$\mathbf{v}_1 = (0,0,0), \quad \mathbf{v}_2 = (1,0,0), \quad \mathbf{v}_3 = (0,1,0) \quad \mathbf{v}_4 = (1,1,1).$$

The vertices are points of certainty: they represent extreme cases in which definite events E_i (properties) pertain to the system. The properties E_i that are represented by the vertices $\mathbf{v}_i$ are

$$E_1 = \neg a \wedge \neg b, \quad E_2 = a \wedge \neg b, \quad E_3 = \neg a \wedge b, \quad E_4 = a \wedge b$$

and are the atoms of the Boolean lattice $L_B^{(2)}$ generated by a and b (Fig. A6.2). The inverse logical equivalence relations

$$a = E_2 \vee E_4, \quad b = E_3 \vee E_4 \quad a \wedge b = E_4$$

can be used to express the properties $a, b, a \wedge b$ in terms of the atoms E_i.

A probability $\mathbf{p} = (p_1, p_2, p_{12})$ that refers to properties of a system must be representable as a convex combination $\mathbf{p} = \sum_{i=1}^{4} \lambda_i(E_i)\mathbf{v}_i$ of the vertex properties E_i with $\lambda_i \geq 0$ and $\sum_{i=1}^{4} \lambda_i = 1$. The coefficients $\lambda_i(E_i)$ are the probabilities of the vertex properties E_i. According to the equivalence relations between the atomic properties E_i and the original properties $a, b, a \wedge b$

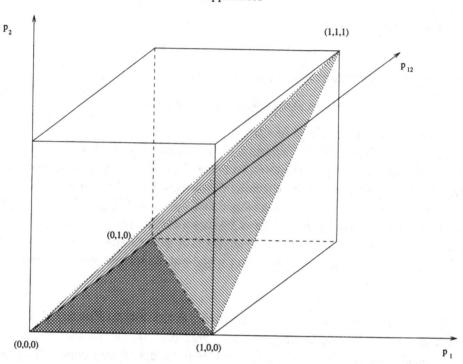

Fig. A6.1 The Boolean polytope $c(2)$.

the respective probabilities $\lambda_i(E_i)$ and $p_1 = p(a)$, $p_2 = p(b)$, $p_{12} = p(a \wedge b)$ are connected by the equations

$$p_1 = \lambda_1 + \lambda_4, \quad p_2 = \lambda_3 + \lambda_4, \quad p_{12} = \lambda_4.$$

The representation of **p** by the vertices $\mathbf{v}_i$ is thus

$$\mathbf{p} = (p_1, p_2, p_{12}) = \lambda_1 \mathbf{v}_1 + \lambda_2 \mathbf{v}_2 + \lambda_3 \mathbf{v}_3 + \lambda_4 \mathbf{v}_4$$
$$= (1 - p_1 - p_2 + p_{12})\mathbf{v}_1 + (p_1 - p_{12})\mathbf{v}_2 + (p_2 - p_{12})\mathbf{v}_3 + p_{12}\mathbf{v}_4.$$

It is obvious that the λ_i fulfill the convexity conditions $\lambda_i \geq 0$, $\sum \lambda_i = 1$, whenever the conditions (4.42) hold. Hence the classicality conditions (4.42) are also *sufficient* for probability attribution.
∎

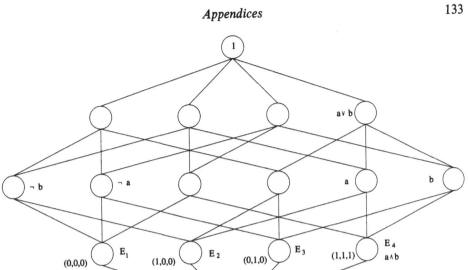

Fig. A6.2 The Boolean lattice $L_B^{(2)}$.

References

[Ama 88] Amann, A. (1988), Chirality as a classical variable in algebraic quantum physics, in *Fractals, Quasicrystals, Chaos, Knots and Algebraic Quantum Mechanics*, eds. A. Amann, L. Cederbaum and W. Gans, Kluever, Dortrecht, pp. 305–25.

[Asp 82] Aspect, A., P. Grangier and G. Rogier (1982), Experimental tests of realistic local theories via Bell's theorem, *Phys. Rev. Lett.* **47**, pp. 460–3.

[BeCa 81] Beltrametti, E. and G. Cassinelli (1981), *The Logic of Quantum Mechanics*, Addison-Wesley, Reading MA.

[Bell 64] Bell, J. S. (1964), On the Einstein–Podolsy–Rosen paradox, *Physics* **1**, pp. 195–200.

[BeMa 91] Beltrametti, E. G. and M. J. Maczynski (1991), On a characterization of classical and nonclassical probabilities, *J. Math. Phys.* **32**, pp. 1280–6.

[BGL 95] Busch, P., M. Grabowski and P. Lahti (1995), *Operational Quantum Physics*, Springer, Heidelberg.

[BHJ 26] Born, M., W. Heisenberg and P. Jordan (1926), Zur Quantenmechanik II, *Zeit. Phys.* **35**, pp. 557–615.

[BLM 91] Busch, P., P. Lahti and P. Mittelstaedt (1991), *The Quantum Theory of Measurement*, Springer, Heidelberg (2nd edition 1996).

[BLM 92] Busch, P., P. Lahti and P. Mittelstaedt (1992), Weak objectification, joint probabilities, and Bell inequalities in quantum mechanics, *Found. Phys.* **22**, pp. 949–62.

[BKLM 92] Busch, P., P. Kienzler, P. Lahti and P. Mittelstaedt (1993), Testing quantum mechanics against a full set of Bell inequalities, *Phys. Rev.* **A 47**, pp. 4627–31.

[BoAh 57] Bohm, D. and Y. Aharonov (1957), Discussion of experimental proof for the paradox of Einstein, Rosen and Podolsky, *Phys. Rev.* **108**, pp. 1070–6.

[Bohr 28] Bohr, N. (1928), The quantum postulate and the recent development of atomic theory, in *Atti del Congresso Internationale del Fisici, Como, 11–20 Septembre 1927*, Zanichelli, Bologna, pp. 565–8.

[Bohr 48] Bohr, N. (1948), On the notion of causality and complementarity, *Dialectica* **2**, pp. 312–19.

[Bohr 49] Bohr, N. (1949), Discussion with Einstein on epistemological problems in atomic physics, in *Albert Einstein: Philosopher–Scientist*, ed. P. A. Schilpp, The Library of Living Philosophers, Tudor Publ. Co.

[BoJo 26] Born, M. and P. Jordan (1926), Zur Quantenmechanik, *Zeit. Phys.* **34**, pp. 858–88.

[Boole 1854] Boole, G. (1854), *The Laws of Thought*, Dover Edition, New York, 1958.

[Boole 1862] Boole, G. (1862), On the theory of probabilities, *Phil. Trans. Roy. Soc. (London)* **152**, pp. 225–52.

[Born 26] Born, M. (1926), Zur Quantenmechanik der Stoßvorgänge, *Zeit. Phys.* **37**, pp. 863–7.

[Breu 94] Breuer, Th. (1994), Classical observables, measurement and quantum mechanics, Ph.D. thesis, University of Cambridge.

[Breu 96] Breuer, Th. (1996), *Quantenmechanik – Ein Fall für Gödel?*, Spektrum, Heidelberg.

[Bub 92] Bub, J. (1992), Quantum mechanics without the projection postulate, *Found. Phys.* **22**, pp. 737–54.

[Bub 94] Bub, J. (1994), On the structure of quantal propositional systems, *Found. Phys.* **24**, pp. 1261–79.

[Bus 85] Busch, P. (1985), Elements of unsharp reality in the EPR experiment, in *Symposium on the Foundation of Modern Physics*, eds. P. Lahti and P. Mittelstaedt, World Scientific, pp. 343–57.

[Bus 96] Busch, P. (1996), Can 'unsharp objectification' solve the quantum measurement problem?, preprint, to be published in *Int. Journ. Theor. Phys.*

[BuS 89] Busch, P. and F. E. Schroeck (1989), On the reality of spin and helicity, *Found. Phys.* **19**, pp. 807–72.

[BuShi 96] Busch, P. and A. Shimony, (1996), Insolubility of the quantum measurement problem for unsharp observables, in *Studies in History and Philosophy of Modern Physics*, to appear.

[CaLa 93] Cassinelli, G. and P. Lahti (1993), The Copenhagen variant of the modal interpretation and quantum theory of measurement, *Found. Phys. Lett.* **6**, pp. 533–44.

[CaLa 95] Cassinelli, G. and P. Lahti (1995), Quantum theory of measurement and the modal interpretation of quantum mechanics, *Int. Journ. Theor. Phys.* **34**, pp. 1271–81.

[DaCh 77] Dalla Chiara, M. L. (1977), Logical self-reference, set theoretical paradoxes, and the measurement problem in quantum mechanics, *Journ. Philos. Logic* **6**, pp. 331–47.

[DeW 70] DeWitt, B. S. (1970), Quantum mechanics and reality, *Physics Today* **23**, pp. 30–5.

[DeW 71] DeWitt, B. S. (1971), The many-universes interpretation of quantum mechanics, in *Foundations of Quantum Mechanics, Proceedings of the International School of Physics 'Enrico Fermi', Course IL, Varenna on Lake Como*, ed. B. d'Espagnat, Academic Press, New York, pp. 167–218.

[DeWG 73] B. S. DeWitt and N. Graham, eds., *The Many-Worlds Interpretation of Quantum Mechanics*, Princeton University Press, Princeton NY (1973).

[Die 89] Dieks, D. (1989), Quantum mechanics without the projection postulate and its realistic interpretation, *Found. Phys.* **19**, pp. 1395–423.

[Die 93] Dieks, D. (1993), The modal interpretation of quantum mechanics, measurements and macroscopic behaviour, *Phys. Rev.* **A 49**, pp. 2290-9.

[Dir 26] Dirac, P. A. M. (1926), On quantum algebra, *Proc. Cam. Philos. Soc.* **23**, pp. 412-18.

[EPR 35] Einstein, A., B. Podolsky and N. Rosen (1935), Can quantum-mechanical description of physical reality be considered complete?, *Phys. Rev.* **47**, pp. 777–80.

[EPS 72] Ehlers, J., F. A. E. Pirani and A. Schild (1972), The geometry of free fall and light propagation, in *General Relativity*, ed. L. O'Raiffeartaigh, Clarendon, Oxford, pp. 63–84.

[Eve 57] Everett, H. (1957), The theory of the universal wave function, Ph.D. thesis, Princeton University, reprinted in *The Many Worlds Interpretation of Quantum Mechanics*, eds. B.S. DeWitt and N. Graham, Princeton University Press, Princeton NY (1973), pp. 1–140.

[Fine 82] Fine, A. (1982), Hidden variables, joint probability and Bell inequalities, *Phys. Rev. Lett.* **48**, pp. 291–5.

[Fink 62] Finkelstein, D. (1962), The logic of quantum physics, *Trans. New York Acad. Sci.* **25**, pp. 621–37.

[Fraa 90] van Fraassen, B. C. (1990), The problem of measurement in quantum mechanics, in *Symposium on the Foundation of Modern Physics 1990*, eds. P. Lahti and P. Mittelstaedt, World Scientific, pp. 497–503.

[Fraa 91] van Fraassen, B. C. (1991), *Quantum Mechanics: An Empiricist View*, Clarendon Press, Oxford, 1991.

[GeHa 90] Gell-Mann, M. and J. B. Hartle (1990), Quantum mechanics in the light of quantum cosmology, in *Proc. 3rd. Int. Symp. Foundations of Quantum Mechanics 1989*, The Physical Society of Japan, pp. 321–43.

[Gie 73] Giere, R. (1973), Objective single-case probabilities, in *Logic, Methodology and Philosophy of Science*, ed. P. Suppes *et al.*, North-Holland, Amsterdam, pp. 467–83.

[Gis 89] Gisin, N. (1989), Stochastic quantum dynamics and relativity, *Helvetica Physica Acta* **62**, pp. 363–71.

[Gis 90] Gisin, N. (1990), Weinberg's non-linear quantum mechanics and superluminal communication, *Phys. Lett.* **A 143**, pp. 1–2.

[Göd 31] Gödel, K. (1931), Über formal unentscheidbare Sätze der Principia Mathematica und verwandter Systeme I, *Monatshefte für Mathematik und Physik* **38**, pp. 173–98.

[Gra 70] Graham, N. (1970), The Everett interpretation of quantum mechanics, Ph.D. thesis, University of Carolina.

[Gra 73] Graham, N. (1973), The measurement of relative frequency, in *The Many-Worlds Interpretation of Quantum Mechanics*, eds. B. S. DeWitt and N. Graham, Princeton University Press, Princeton NY ([DeWG 73]), pp. 229–52.

[Gro 90] Gross, H. (1990), Hilbert lattices: New results and unsolved problems, *Found. Phys.* **20**, pp. 529–59.

[GRW 86] Ghirardi, G. C., A. Rimini and T. Weber (1986), Unified dynamics for microscopic and macroscopic systems, *Phys. Rev.* **D 34**, pp. 470–91.

[HaHa 83] Hartle, J. B. and S. W. Hawking (1983), Wave function of the universe, *Phys. Rev.* **D 28**, pp. 2960–75.

[Hart 68] Hartle, J. B. (1968), Quantum mechanics of individual systems, *Am. Journ. Phys.* **36**, pp. 704–12.

[Heis 25] Heisenberg, W. (1925), Über quantentheoretische Umdeutung kinematischer und mechanischer Beziehungen, *Zeit. Phys.* **33**, pp. 879–93.

[Heis 59] Heisenberg, W. (1959), Die Plancksche Entdeckung und die physikalischen Probleme der Atomphysik, *Universitas* **14**, pp. 135–48.

[Heis 69] Heisenberg, W. (1969), *Der Teil und das Ganze*, Pieper & Co., München.

[Hume 1739] Hume, D. (1739/40), *A Treatise of Human Nature*.

[Jam 74] Jammer, M. (1974), *The Philosophy of Quantum Mechanics*, John Wiley & Sons, New York.

[Jau 68] Jauch, J. M. (1968), *Foundations of Quantum Mechanics*, Addison-Wesley, Reading MA.

[Joos 84] Joos, E. (1984), Continuous measurement: Watchdog effect versus golden rule, *Phys. Rev.* **D 29**, pp. 1626–33.

[JoZe 85] Joos, E. and H. D. Zeh (1985), The emergence of classical properties through interaction with the environment, *Zeit. Phys.* **B 59**, pp. 223–43.

[Kant 1787] Kant, I. (1787), *Kritik der reinen Vernunft*, Hartknoch, Riga, English translation *I. Kant's Critique of Pure Reason*, N. K. Smith (1929), MacMillan, London.

[Kel 80] Keller, H. A. (1980), Ein nichtklassischer Hilbertscher Raum, *Math. Zeit.* **173**, pp. 41–9.

[Koch 85] Kochen, S. (1985), A new interpretation of quantum mechanics, in *Symposium on Foundations of Modern Physics*, eds. P. Lahti and P. Mittelstaedt, World Scientific, Singapore, pp. 151–69.

[KuHo 62] Kundt, W. and B. Hofman (1962), Determination of gravitational standard time, in *Recent Developments in General Relativity*, Warschau, PWN, pp. 303–6.

[Mach 26] Mach, E. (1926), *Erkenntnis und Irrtum*, 5th edition, Barth, Leipzig.

[Mack 63] Mackey, G. W. (1963), *Mathematical Foundations of Quantum Mechanics*, Benjamin, New York.

[MaWh 64] Marzke, R. F. and J. A. Wheeler (1964), Gravitation as geometry I: The geometry of space–time and the geometrodynamical standard meter, in *Gravitation and Relativity*, ed. H. Y. Chin, Benjamin, New York, pp. 40–64.

[Mit 78] Mittelstaedt, P. (1978), *Quantum Logic*, Reidel, Dortrecht, Holland.

[Mit 86] Mittelstaedt, P. (1986), *Sprache und Realität in der modernen Physik*, BI Wissenschaftsverlag, Mannheim.

[Mit 93] Mittelstaedt, P. (1993), The measuring process and the interpretation of quantum mechanics, *Int. Journ. Theor. Phys.* **32**, pp. 1763–75.

[Mit 95] Mittelstaedt, P. (1995), The constitution of objects in classical mechanics and in quantum mechanics, *Int. Journ. Theor. Phys.* **34**, pp. 1615–26.

[MPS 87] Mittelstaedt, P., P. Prieur and R. Schieder (1987), Unsharp particle–wave duality in a photon split-beam experiment, *Found. Phys.* **17**, pp. 891–903.

[Mun 93] Mundice, D. (1993), Logic of infinite quantum systems, *Int. Journ. Theor. Phys.* **32**, pp. 1941–55.

[Neu 32] von Neumann, J. (1932), *Mathematische Grundlagen der Quantenmechanik*, Springer, Berlin.

[Oza 84] Ozawa, M. (1984), Quantum measuring process of continuous observables, *Journ. Math. Phys.* **25**, pp. 79–87.

[Pea 86] Pearle, P. M. (1986), Stochastic dynamical reduction theories and superluminal communication, *Phys. Rev.* **D 33**, pp. 2240–52.

[Pen 58] Penrose, L. S. and R. Penrose (1958), Impossible objects: A special type of visual illusion, *Brit. Journ. Psych.* **49**, pp. 1–3.

[PeZu 82] Peres, A. and W. Zurek (1982), Is quantum-theory universally valid?, *Am. Journ. Phys.* **50**, pp. 807–10.

[Pfei 80] Pfeifer, P. (1980), Chiral molecules – a superselection rule induced by the radiation field, Ph.D. thesis, ETH Zürich, No. 6551.

[Pir 76] Piron, C. (1976), *Foundations of Quantum Physics*, Benjamin, Reading MA.

[Pit 89] Pitowsky, I. (1989), *Quantum Probability – Quantum Logic*, Lecture Notes in Physics, Vol. 321, Springer, Berlin.

[Pit 94] Pitowsky, I. (1994), George Boole's 'conditions of possible experience' and the quantum puzzle, *Brit. Journ. Philos. Sci.* **45**, pp. 95–125.

[Pop 57] Popper, R. K. (1957), The propensity interpretation of the calculus of probability and the quantum theory, in *Observation and Interpretation in the Philosophy of Physics*, ed. S. Körner, Butterworth, London.

[Prim 83] Primas, H. (1983), *Chemistry, Quantum Mechanics and Reductionism*, Springer, Berlin.

[Pop 82] Popper, R. K. (1982), Quantum theory and the schism of physics, in *Postscript III to the Logic of Scientific Discovery*, ed. W. W. Bartley III, Hutchinson, London, Melbourne, 1988.

[Schm 07] Schmidt, E. (1907), Zur Theorie de linearen und nichtlinearen Integralgleichungen, I. Teil: Entwicklung willkürlicher Funktionen nach Systemen vorgeschriebener, *Math. Ann.* **63**, pp. 433–76.

[Schn 94] Schneider, C. (1994), Two interpretations of objective probabilities, *Philosophia Naturalis* **31**, pp. 107–31.

[Schrö 26] Schrödinger, E. (1926), Quantisierung als Eigenwertproblem, *Annalen der Physik* **79**, pp. 361–76.

[Shim 89] Shimony, A. (1989), Search for a worldview which can accommodate our knowledge of microphysics, in *Philosophical Consequences of Quantum Theory; Reflections on Bell's Theorem*, eds. J. T. Cushing and E. McMullin, The University of Notre Dame Press.

[Sim 78] Simonius, M. (1978), Spontaneous symmetry breaking and blocking of metastable states, *Phys. Rev. Lett.* **40**, pp. 980–3.

[Skl 70] Sklar, L. (1970), Is probability a dispositional property?, *Journ. Philos.* **67**, pp. 355–66.

[Sol 95] Solèr, M. P. (1995), Characterization of Hilbert spaces by orthomodular spaces, preprint.

[Sta 80] Stachow, E. W. (1980), Logical foundation of quantum mechanics, *Int. Journ. Theor. Phys.* **19**, pp. 251–304.

[Wei 89] Weinberg, S. (1989), Testing quantum mechanics, *Ann. Phys.* **194**, pp. 336–86.

[Whe 57] Wheeler, J. A. (1957), Assessment of Everett's 'relative state' formulation of quantum theory, *Rev. Mod. Phys.* **29**, pp. 463–5.

[Wig 63] Wigner, E. (1963), The problem of measurement, *Am. Journ. Phys.* **31**, pp. 6–15.

[Zeh 70] Zeh, D. (1970), On the interpretation of measurement in quantum theory, *Found. Phys.* **1**, pp. 69–76.

[Zur 82] Zurek, W. H. (1982), Environment induced superselection rules, *Phys. Rev.* **D 26**, pp. 1862–80.

Index

Index

Printed in the United States
By Bookmasters